THE MACROPHAGE

B. VERNON-ROBERTS

M.D., PH.D.

Senior Lecturer in Morbid Anatomy
The London Hospital Medical College
University of London

CAMBRIDGE
AT THE UNIVERSITY PRESS
1972

Published by the Syndics of the Cambridge University Press
Bentley House, 200 Euston Road, London NW1 2DB
American Branch: 32 East 57th Street, New York, N.Y.10022

© Cambridge University Press 1972

Library of Congress Catalogue Card Number: 72-184141

ISBN: 0 521 08481 4

Printed in Great Britain
at the University Printing House, Cambridge
(Brooke Crutchley, University Printer)

CONTENTS

To Jayne
to whose help and encouragement
this book owes its existence

PREFACE

The macrophages, which are widely distributed throughout the body tissues, have long been known to play a major role in host defence mechanisms and are the cells ultimately responsible for the disposal of most foreign materials and damaged host tissues. It is widely recognized that they participate in a variety of metabolic processes in both normal and pathological circumstances. In recent years, increasing attention has been paid to these cells as participants in cell-mediated immunological reactions and antibody production. The wide range of activities of macrophages has ensured that their study has continued to engage the interest of a wide range of scientific disciplines.

The purpose of this monograph is to provide an account of present knowledge regarding the life-history of the macrophage, and the evidence for its role in various physiological and pathological processes. I hope that a clear demarcation exists between the presentation of the findings and conclusions of other authors and my own speculations. Where controversial topics are presented, I have consciously endeavoured to limit my approach to the examination of the relevant evidence without bias.

I hope that the scope of this monograph and the approach which I have used will not only appeal to those who have a research interest in this field, but will also be of value to clinicians and pathologists seeking a deeper understanding of basic mechanisms underlying body defence mechanisms and certain disease processes.

I am indebted to Professor T. Nicol for introducing me to the ubiquitous macrophage, and to many other colleagues with whom I have had the pleasure of working or have discussed the many aspects of this cell. It gives me particular pleasure to record my gratitude to Professor I. Doniach for his constant interest and encouragement during the preparation of this monograph.

I thank all those colleagues who have generously allowed me to use illustrations from their published work, and I have pleasure in recording the contributions of Dr I. Carr; Professor B. N. Halpern, Dr G. Biozzi and Dr B. Benacerraf; Dr M. G. Hanna and Dr A. K. Szakel; Dr A. H. E. Marshall; Dr R. J. North; and Professor T. Nicol, Dr D. L. J. Bilbey, Dr J. L. Cordingley and Dr L. M. Charles. I thank the editors and publishers who granted me permission to use these illustrations, the sources of which are acknowledged in the legends for the relevant figures, and the original place of publication recorded in the references.

I am particularly indebted to Mrs Lynda Moore for the patience and care with which she typed out draft and final copies of the manuscript, to Mr R. M. Hammond and Mr G. Walter for preparation of most of the illustrations, and to Mr A. Gray for preparing the electron micrographs.

Grateful acknowledgements are due to the librarians of The London Hospital Medical College, The British Medical Association, The University of London Library at Senate House, and The National Lending Library.

Last, but not least, I record my gratitude to my wife. Not only has she relieved me of all domestic problems and spent many hours of solitude during the prolonged gestation of the manuscript, but she also assisted in checking the typescript and reading the proofs. Without her whole-hearted and constant encouragement, the task of preparing this monograph would have been beyond my powers of endurance.

B. VERNON-ROBERTS

INTRODUCTION

The presence of amoeboid mononuclear cells at inflammatory sites appears to have been first described by von Recklinghausen in 1863, after he had observed them in the inflamed cornea and omentum in various species of animals. He was able to distinguish these amoeboid cells from pus cells (polymorphonuclear leucocytes) and postulated that they may have been derived from fixed connective tissue cells. Some twenty years later, Metchnikoff drew attention to the fact that in addition to the free mononuclear phagocytic cells of the blood and lymph there were normally present in the connective tissues and certain organs fixed cells which were able to engulf small particles by throwing out amoeboid processes. In an outstanding series of publications, Metchnikoff went on to describe the distribution of these cells in the liver, spleen, lymph nodes and in the central nervous system in invertebrates and vertebrates, including man. He later grouped the free and fixed large mononuclear cells together as 'macrophages', and distinguished them from the leucocytes of the circulating blood which he called 'microphages' (Metchnikoff, 1905).

Although it had been recognized for many years prior to Metchnikoff's observations that micro-organisms were commonly found within white cells during inflammation, the opinion had been generally held that micro-organisms found the interior of leucocytes to be favourable to their survival, and it was generally believed that leucocytes played a role in disseminating micro-organisms throughout the body. However, the observation that leucocytes emigrated into the tissues during inflammation suggested to Metchnikoff that these cells must play a protective role in host defence. He set about proving his hypothesis by inserting rose prickles into the transparent bodies of the larvae of starfish, and observed the accumulation of amoeboid cells around the foreign body, noting the similarity to inflammation induced in the human. He extended his studies by introducing various kinds of bacteria into lower animals, and demonstrated that they were ingested by similar cells. By clearly demonstrating the defensive role of phagocytic cells, Metchnikoff established the cellular theory of body defence.

The study of macrophage morphology and distribution was much enhanced by the introduction of non-toxic vital dyes suitable for histological use. These dyes, such as trypan red and trypan blue, are preferentially ingested by macrophages. Using vital staining techniques it was subsequently shown by many workers that macrophages exist as 'free' cells scattered extravascularly throughout the connective tissues and among the cells of the thymus, spleen and lymph nodes; and also as 'fixed' sessile cells lying along the walls of blood sinusoids in the liver (Kupffer cells), spleen, bone marrow and other sites, and along the walls of lymph sinuses in lymph nodes. Different investigators often gave special names to these cells to designate either their function or their presumed origin, and the macrophages have been synonymously called clasmatocytes, rhagiocrine cells, polyblasts, reticulum cells, reticular cells, polymorphous histogenic wandering cells, pyrrhol cells, resting wandering cells and histiocytes.

Aschoff was the first to recognize that the mononuclear phagocytes probably constituted a functionally unified system of cells, and later introduced the term 'reticulo-endothelial system' to cover the entire range of cells which possessed the capacity to take up vital dyes (Aschoff, 1924). The term 'reticulo-endothelial system' was derived from the fact that the cells forming the system were considered to be involved in the formation of the 'reticulum' (extrasinusoidal solid pulp) of the lymph nodes and spleen, or were those lining blood or lymph sinusoids. Aschoff divided all cells which have the capacity for ingesting vital dyes into four groups. Cells of groups I and II exhibited relatively little vital staining capacity and consisted of the endothelium of blood and lymphatic vessels and the fibroblasts of the connective tissues. Group III consisted of the 'reticulum' (pulp) cells of spleen or lymph nodes. Group IV comprised the sinusoidal lining cells of lymph node sinuses, and the sinusoids of the liver, spleen, bone marrow, adrenal cortex and pituitary. This latter group also included the histiocytes (free macrophages) of extravascular connective tissues. Cells of groups III and IV were considered by Aschoff to form the 'reticulo-endothelial system', while those of groups I and II were excluded.

About the same time that Aschoff was delineating his 'reticulo-endothelial system', Del Rio Hortega and dé Asua (1924) demonstrated the presence of the microglial cells in the central nervous system by impregnating them with silver carbonate. In cerebral inflammation and injury the microglial cells could be shown to become transformed into the characteristic mononuclear phagocytes of the central nervous system – the 'compound granular corpuscles'. Because of the resemblance between these phagocytes of the central nervous system and those elsewhere in the body, Hortega and dé Asua applied the silver carbonate technique to various tissues outside the central nervous system and showed that cells similar to the microglia were present in them, and that these cells had phagocytic properties.

Subsequent workers showed that silver impregnation not only demonstrated a system of cells which included the 'reticulo-endothelial system' as defined by Aschoff, but also included a related series of phagocytic cells which were not demonstrable by vital staining. The cytoplasm of all these cells had such an affinity for metal salts that they were called 'metalophil cells'. Marshall (1956) made a detailed and critical survey of the morphology and distribution of the metalophil cells and resolved a number of discrepancies in the results of vital staining and in terminology. He concluded that there is no exact parallelism between the metalophil cells and the 'reticulo-endothelial system' as defined by Aschoff, and that the metalophil cells are considerably more numerous. It thus appears that the 'reticulo-endothelial system' is not a distinct cytological entity but is a functional state of some of the metalophilic macrophages depending largely on local factors such as cell disposition and lymph and blood circulation. This is clearly exemplified by the blood monocytes which are metalophilic but do not take up vital dyes. Moreover, using a silver impregnation technique, Marshall was able to distinguish and describe the different morphology of the strongly metalophilic macrophages and reticulum cells and the primitive reticular cells which were not metalophil.

Although the mononuclear phagocytes of the body represent a large, widely distributed and morphologically somewhat heterogeneous group of cells, it is now evident that, as described in later chapters of this monograph, the macrophages have ultrastructural, metabolic and functional characteristics which may be used to distinguish them from other cells. It would appear that to avoid confusion the simplest and best appellation to include all the macrophages of the body seems to be 'the macrophage system'.

The most striking functional characteristic of macrophages is phagocytosis. At one time, the uptake of both solid particles and fluid droplets were considered to come under the heading of phagocytosis. However, Lewis (1931) introduced the term 'pinocytosis' (drinking by cells) to name the process by which macrophages take up fluid droplets. The electron microscope has revealed that a variety of cells other than macrophages can also take up submicroscopic particles and fluid droplets in vesicles formed by invagination of the cell membrane. It would appear that the mechanism of particle uptake (phagocytosis) and fluid uptake (pinocytosis) may in fact be identical and differ only in the content of vesicle formed, and the term 'endocytosis' is often used to encompass both processes.

Although the properties of phagocytosis and pinocytosis are not confined to macrophages, this does not invalidate the specificity of the macrophage as a cell which is able to recognize and ingest particles which may be relatively large and which also may be harmful to the organism (bacteria, fungi, yeasts, protozoa, effete erythrocytes, dead leucocytes, foreign bodies, etc.). It is now known that the macrophage system is also involved in immune processes, the inactivation of bacterial endotoxins, resistance to traumatic and haemorrhagic shock, the prevention of irradiation sickness, haematoclasia and bile pigment formation, and the metabolism of lipids, steroids, cholesterol, iron and proteins. Moreover, macrophages have been implicated in the aetiology of atherosclerosis, hypercholesterolaemia, lipoidoses, haemolytic anaemias, amyloidosis and neoplasia.

The widespread distribution and the involvement of macrophages in a multiplicity of physiological and pathological processes ensures that the study of the macrophage system of cells is an exercise which can involve many scientific disciplines and can be approached at various levels of interest in all branches of science.

TABLE 1. *The distribution of macrophages in mammals*

Anatomical site	Localization of cells
Liver	Kupffer cells lining hepatic sinusoids; connective tissues of portal tracts
Spleen	Lining venous sinuses and enmeshed in Billroth cords of red pulp; scattered among lymphocytes of Malpighian follicles in white pulp; in marginal zones
Lymph nodes	Lining subcapsular and medullary sinuses; scattered in medullary pulp and in lymphoid follicles of cortex
Bone marrow	Lining venous sinuses of red marrow; scattered monocytes and macrophages in extrasinusoidal tissues
Thymus	Scattered throughout cortex and medulla; within Hassall's corpuscles in some species
Lung	Within interstitial tissue of alveolar wall; in alveolar spaces
Central nervous system	Microglia
Serous cavities	Peritoneal and pleural fluids; 'milk spots' of peritoneum and pleura
Adrenals	Lining sinusoids of cortex (particularly zona reticularis); scattered in medulla
Joints	Type A or M cells of synovial lining; within synovial fluid
Subcutaneous tissue, alimentary tract, pituitary, testis, ovary, endometrium, kidney	Connective tissues generally, lining vascular channels in pituitary, testis, ovary and decidua
Blood	Monocytes and some macrophages
All sites	Inflammatory exudates

1

THE DISTRIBUTION AND MORPHOLOGY
OF MACROPHAGES

ANATOMY OF THE MACROPHAGE SYSTEM

The widespread distribution of the component cells of the macrophage system results in certain morphological differences which are due to the structure, metabolism and blood supply of the organs of which these cells form a part, to their functional requirements, to their relation to tissue fluids, and to their relation to the specific parenchymatous cells of the organ concerned.

For descriptive and functional purposes it is still sometimes profitable broadly to follow Metchnikoff in considering that the macrophages of the body may be grouped as 'fixed' macrophages lying along the walls of blood and lymph channels in direct contact with circulating blood or lymph, and 'free' macrophages situated extravascularly in the connective tissues, in solid organs, and in the central nervous system. However, it is important to remember that both the 'fixed' and 'free' cells can be mobilized, and that these prefixes are used for descriptive convenience alone. Moreover, there are additional component cells of the system which are present in the circulating blood, and comprise the monocytes (immature macrophages) and mature macrophages.

Table 1 summarizes the anatomical distribution of the cells which are considered as members of the macrophage system in mammals.

MACROPHAGES IN SPECIAL SITES

LIVER

The liver contains more macrophages than any other single organ. Light microscopic studies after metallic impregnation, vital staining or intravenous injection of colloids reveal that the major portion of each liver sinusoid is lined by actively phagocytic branching stellate macrophages, the Kupffer cells, whereas the central and peripheral regions of each sinusoid are lined by less actively phagocytic endothelial-type cells. A considerable number of studies on liver ultrastructure have been carried

[5

out, and there is general agreement that electron microscopy confirms the presence of two types of cell, the endothelial cell and the Kupffer cell, lining the sinusoids. However, there are conflicting reports regarding the absence or presence of a true basement membrane underlying the cells lining the sinusoids, and also with regard to the existence of gaps between adjacent cells; and from the available evidence, it appears that the structure of the liver sinusoid may vary from one species to another (Burkel and Low, 1965). In general, it seems that the endothelial lining of the peripheral part of the sinusoid is continuous and has an underlying basement membrane: the part of the sinusoid adjacent to the central vein also has a continuous endothelial lining with an underlying basement membrane; between the central and peripheral portions (comprising about 90 per cent of the length of the sinusoid) gaps are present between adjacent cells; Kupffer cells are present in this intermediate portion of the sinusoid and do not have an underlying basement membrane.

The endothelial cells and Kupffer cells are separated from the hepatocytes (liver parenchymal cells) by the space of Disse (Fig. 1). The absorptive surfaces of the hepatocytes are differentiated into short microvilli which occupy the space of Disse together with a network of delicate collagen fibrils. Cells described by some authors as pericytes may be present, but some undoubtedly have the fine structure of macrophages. Occasional cells rich in lipid inclusions, and undifferentiated mesenchymal cells known as 'Disse-space cells', are also occasionally observed within the space of Disse (the perisinusoidal space).

When normal untreated animals are sacrificed after the intravenous injection of a single dose of colloidal carbon, the liver exhibits phagocytosed carbon located within Kupffer cells located in the mid-zonal and peripheral portions of the liver lobules (Fig. 2). After repeated injections of carbon, or pre-treatment of the animals with oestrogen, it can be seen that all the cells lining the liver sinusoids take up carbon (Fig. 3). It would thus seem that the endothelial cells may be 'recruited' to act as phagocytes, since they are capable of displaying phagocytic activity before the proliferation of Kupffer cells or the influx of macrophages from elsewhere can occur. This suggests that the endothelial cells of the liver sinusoids differ from those lining vascular channels in most areas of the body, since the latter do not exhibit any marked degree of phagocytosis. It would seem reasonable, therefore, to regard the liver sinusoids as being lined by a mixed population of true mature macrophages (the Kupffer cells) and specialized 'endothelioid' or potential macrophages. In this connection, although the endothelial cells normally have markedly different ultrastructural appearances from Kupffer cells, they do not have all the features of true capillary endothelium. Thus, their fenestrations differ from those of true capillary endothelium by the absence of a

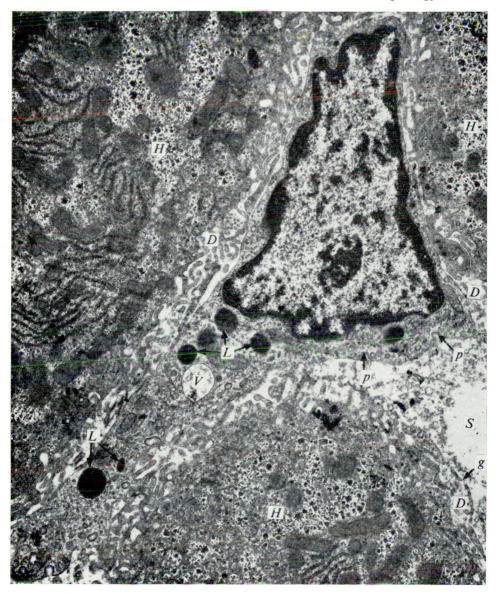

Fig. 1. Kupffer cell in guinea-pig. The macrophage contains several dense bodies (*L*) which are probably lysosomes, a vacuole (*V*) which is probably phagocytic in origin, and exhibits pinocytotic activity (*p*). It is separated from the hepatocytes (*H*) by the space of Disse (*D*) which contains the microvilli of the hepatocytes and communicates with the lumen of the sinusoid (*S*) via inter-cellular gaps (*g*). × 12,500.

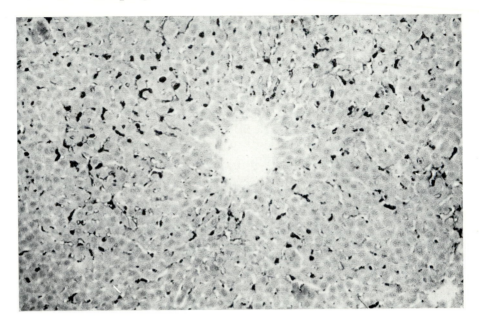

Fig. 2. Liver of untreated guinea-pig 10 minutes after intravenous injection of
carbon. Shows carbon-containing Kupffer cells located in intermediate portion
of liver sinusoids. Neutral red. × 120.

diaphragm within their fenestrations, and they do not consistently
possess an underlying basement membrane (Wisse, 1970). The Kupffer
cells do not possess fenestrations, and usually contain prominent lysosomes
and occasional phagocytic vacuoles (Fig. 1).

The liver also contains some macrophages which are located in the
extravascular connective tissues of the portal tracts.

SPLEEN

The spleen contains more macrophages per gram of tissue than any other
organ, and they are situated predominantly in the red pulp and the
marginal zones, together with small numbers scattered within the Malpighian bodies comprising the white pulp (Fig. 4).

It would appear that the blood passing through the spleen is effectively
'filtered' by exposure to macrophages within the red pulp. The venous
sinuses are lined by littoral macrophages (macrophages which form the
walls of blood channels) which are supported by reticulin fibres forming
an incomplete membrane, and the striking feature of the sinuses is the
large gaps which exist between the littoral cells and which appear to

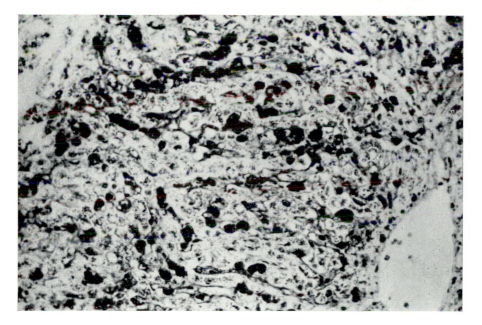

Fig. 3. Liver of oestrogen-treated guinea-pig 10 minutes after intravenous injection of carbon. There has been heavy carbon uptake by cells throughout the lining of the sinusoids. Neutral red. × 120.

allow the free and unhindered passage of erythrocytes into the red pulp. In the Billroth cords of the red pulp, macrophages are embedded in a fine reticulin meshwork, and for the most part form a syncitial network of interconnecting cell processes. Free rounded amoeboid macrophages are seen in the venous sinuses and in the cords of the red pulp. At the junction of red and white pulp is the marginal zone, a capacious irregular sinus, intersected by numerous reticular cells, communicating with both the capillaries of the white pulp and the cords and sinuses of the red pulp, and which is packed with macrophages and other leucocytes (Carr, 1970). In the white pulp, a few macrophages having dendritic processes are normally scattered among the lymphoid cells of the Malpighian bodies. Other cells, the 'antigen-retaining reticular cells', which may be variants of the macrophage, form an extensive web-like meshwork of fine interconnecting processes in the lymphoid follicles during immunological responses, and retain demonstrable antigen on the surfaces of their plasma membranes. The role of these cells in immunological responses is discussed in Chapter 7.

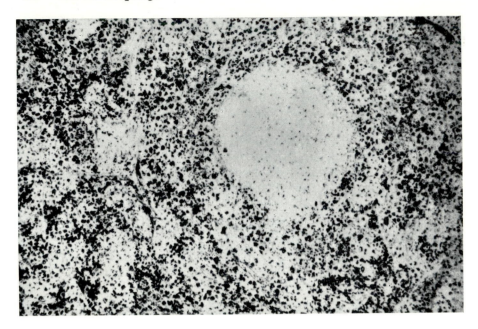

Fig. 4. Spleen of untreated guinea-pig 24 hours after intravenous injection of carbon. Shows abundant carbon-containing macrophages in red pulp and marginal zone, and absence of carbon-containing cells in the Malpighian body. Neutral red. × 120.

LYMPH NODES

There are species variations in the general structure of lymph nodes. In some animals, trabeculae, which are fibrous extensions from the capsular surface of the node, may extend for a variable distance into the substance of the node. Trabeculae are not present or are poorly developed in some species. Some authors hold the view that the capsular and trabecular sides of the sinus walls of lymph nodes are lined by nonphagocytic endothelial cells, while on the sides of sinuses adjacent to lymphoid tissue, the lining is composed of phagocytic cells. Others hold the view that all the cells lining the sinuses are endothelial in type and are not significantly phagocytic. Whatever the terminology used, it seems that the lining cells of the sinuses are readily phagocytic in response to particulate matter (Fig. 5), and as such can be considered to be members of the macrophage family. The cells lining the sinuses rest on a fine network of fibres of reticulin, collagen, and elastin, and the collagen fibres may extend across the sinuses and into the dense intersinusoidal lymphoid tissues. Macrophages may be observed enmeshed within the collagen network of the sinusoids. Considerable doubt exists as to whether a true basement membrane

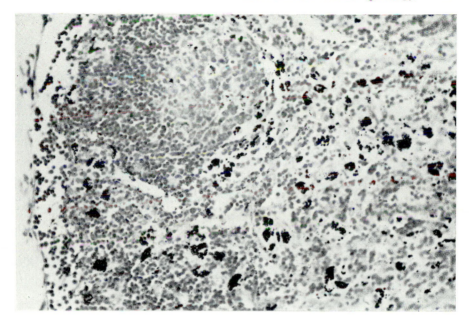

Fig. 5. Lymph node of guinea-pig 24 hours after subcutaneous injection of carbon.
Shows carbon-containing macrophages in sinuses, in cortex adjacent to sinuses,
and in lymphoid follicle. Neutral red. × 120.

underlies the sinusoidal cells, and gaps appear to be present in the
sinusoidal wall.

Scattered macrophages are present in the medullary pulp, and in
lymphoid follicles and parafollicular areas of the cortex (Fig. 5). Some
of the macrophages of the lymphoid follicles are called 'dendritic'
macrophages on account of their cytoplasmic processes which are clearly
demonstrable by metallic impregnation. Other cells, which may be
macrophages, are known as 'antigen-retaining reticular cells' since
antigen becomes bound to their cell membranes. They have ultrastructural
differences from macrophages of other tissues and other areas in spleen
or lymphoid tissues: the cytoplasm of these cells contains numerous
ribosomes, polyribosomes and numerous Golgi-type vesicles; rough
endoplasmic reticulum is not prominent and lysosomes are rare. The
extensive fine processes of adjacent antigen-retaining reticular cells
interdigitate with one another and also with processes of lymphoid cells
during immunological responses (Fig. 47).

Secondary lymphoid follicles and germinal centres contain more
dendritic macrophages and antigen-retaining reticular cells. The macrophages of germinal centres are widely referred to as 'tingible body

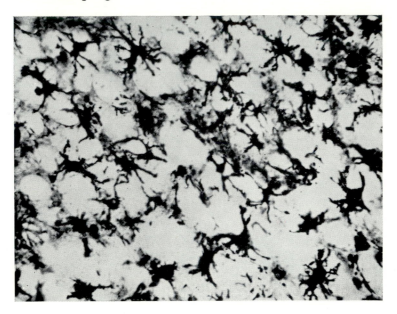

Fig. 6. Fixed metalophil macrophages in normal rabbit bone marrow. Weil–Davenport method. × 220. Reproduced with permission from Marshall (1956).

macrophages' since the cytoplasm consistently contains a quantity of stainable material (Figs. 45, 46). Electron microscopy reveals that they contain a large variety of inclusions and ingested nuclear material in various stages of breakdown which probably represents the ingested debris of lymphocytes or antigen-retaining reticular cells which have died locally.

BONE MARROW

The venous sinuses of the red marrow are lined with flattened macrophages which possess numerous branching and ramifying processes (Fig. 6). Free amoeboid macrophages are also present and frequently contain the debris of erythrocytes and haemosiderin. The bone marrow is the major source of the monocytes which appear at inflammatory sites. The monocytes arise from large phagocytic cells, the promonocytes, which themselves arise from bone marrow precursors which are hitherto unidentified.

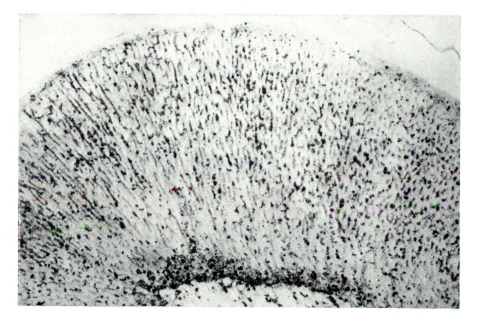

Fig. 7. Adrenal gland of untreated guinea-pig 24 hours after intravenous injection of carbon. Shows carbon-containing macrophages lining sinuses of cortex. Neutral red. × 120.

ADRENALS

Macrophages form part of the lining of blood sinusoids throughout the adrenal cortex, and are particularly prominent in the zona reticularis (Fig. 7). Occasional macrophages are also present in the periendothelial spaces of the cortex. Scattered macrophages are present throughout the blood spaces of the adrenal medulla.

LUNG

Many macrophages are present within the interstitial tissues of the alveolar wall, and within the lumen of the alveolar space either free within the lumen or fixed to the alveolar wall. Occasional macrophages may extend through the interstitium from one alveolar surface to the next, and both surfaces may possess abundant microvilli. Alveolar macrophages may also pass through the pores of Kohn. There is evidence that about one-third of the alveolar macrophages are derived locally, and that two-thirds are derived from precursors situated elsewhere in the body (Pinkett, Cowdrey and Nowell, 1966). It has also been suggested that the normal alveolar macrophage population may be largely derived or augmented from the liver, spleen and connective tissues, and that this represents an excretory pathway for macrophages which have taken up large quantities of material (Nicol and Bilbey, 1960).

For a detailed review of all aspects of lung macrophages, the reader is referred to an excellent account by Stuart (1970).

UTERUS

The cyclical variations in the levels of steroid hormones, known to occur during the reproductive cycle, are accompanied by cyclical alterations in the number of free macrophages in the endometrial stroma. It has been shown in the guinea-pig, rat and mouse that macrophages are present in the endometrial stroma in increased numbers during the follicular phase (at proestrus) and during the luteal phase (at metoestrus) at the time of endometrial breakdown. Moreover, macrophages are rarely seen in the endometrium after ovariectomy, but re-appear following the injection of oestrogen (Nicol, 1935; Nicol and Vernon-Roberts, 1965). A similar cyclical influx of macrophages into the endometrium has also been described in the human (Papanicolaou, 1953).

During pregnancy, there is an influx of macrophages into the decidua (endometrial stroma of pregnancy) about the time of implantation, and towards the end of pregnancy and during the early puerperium in the rat, mouse and guinea-pig (Nicol, 1935; Nicol and Vernon-Roberts, 1965). Using vital staining, Nicol (1935) showed that in the guinea-pig many vitally-stained macrophages are present in the walls of the decidual cavity and around the embedding ovum.

CENTRAL NERVOUS SYSTEM

In the leptomeninges and in the intermeningeal spaces numerous mononuclear amoeboid cells appear under the influence of various irritants; and they display active phagocytosis and behave like macrophages in other parts of the body. The source of these cells is largely unknown. In the choroid plexus, macrophages are less numerous than in the meninges.

In the central nervous system proper, the microglial cells become conspicuous after the injection of dye or carbon directly into the brain; they also become conspicuous when phagocytosing debris after damage to the brain and spinal cord, and in such conditions become loaded with lipoid material and present a foamy appearance in paraffin sections. Under these conditions, they are often called 'compound granular corpuscles'. In the resting condition, the microglial cells have narrow, elongated, irregularly-shaped nuclei and a narrow rim of cytoplasm which extends

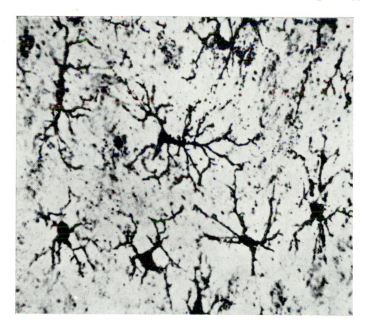

Fig. 8. Extensively branched metalophil microglia in cerebral cortex of human brain. Weil–Davenport method. × 400. Reproduced with permission from Marshall (1956).

into relatively scanty processes characterized by rich ramifications (Fig. 8); but, under the appropriate stimulus, they rapidly transform into free amoeboid forms.

JOINTS

The synovial lining of mammalian joints contains macrophages, often designated as Type A or Type M cells by different workers, which have the electron microscopical appearance of macrophages elsewhere and are actively phagocytic. Moreover, free amoeboid macrophages are present in the synovial fluid.

SEROUS CAVITIES

In the pleural and peritoneal cavities there are abundant macrophages lying free or attached to the walls of these spaces. They occur as large cells with rounded vesicular nuclei, and also as small cells with indented dense nuclei. It appears that the latter type of cell may arise from small lymphocytes of bone marrow origin, and many cells may be observed which have an appearance intermediate between lymphocyte and macrophage (Vernon-Roberts, 1969*a*).

THYMUS

Macrophages are scattered throughout the thymic tissues, and appear to be present within Hassal's corpuscles in some species.

BLOOD

The monocyte can be considered an immature member of the macrophage system, but nevertheless possesses some capacity for phagocytosis in its circulating form. It has been well established that monocytes have the capacity of transforming into mature macrophages. Mature macrophages are also seen in the blood under normal and pathological circumstances.

CONNECTIVE TISSUES GENERALLY

The macrophages of the connective tissues present a variety of appearances and occur in sessile or amoeboid forms. The sessile forms are finely branching cells anchored between the connective tissue fluids. Under normal and pathological circumstances they may withdraw their processes and start migration as amoeboid macrophages. The amoeboid forms are spheroidal and project rounded pseudopodia.

In the omentum, peritoneum and pleura they may be so numerous and closely grouped together that they can be seen with the naked eye as whitish spots, the so-called 'milk spots'.

MACROPHAGES IN OTHER TISSUES

In the pituitary, ovary and testis, macrophages are distributed along blood and lymph channels in these organs, and are also present as scattered clumps in the connective tissues. In the kidney, small numbers of macrophages are present in the interstitial tissues of cortex and medulla. The mesangial cells of the glomerular tufts are capable of displaying active phagocytosis, and after the intravenous injection of colloidal carbon, they may be seen to contain large quantities of the colloid. Similarly, Schwann cells may display very active phagocytosis following damage to nerve fibres. At present, there is not enough evidence to support the inclusion of either mesangial cells or Schwann cells within the system of macrophages.

Variable numbers of monocytes and macrophages are a constant feature of inflammatory sites in any part of the body.

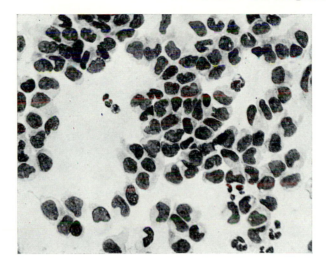

Fig. 9. Monolayer of human macrophages which have migrated on to a glass cover-slip placed over a scratched area of skin. Giemsa. × 1,000.

MORPHOLOGY OF MACROPHAGES

LIGHT MICROSCOPY

Macrophages are morphologically heterogeneous. They vary greatly in size (10–50 μ) and present a variety of nuclear and cytoplasmic appearances. The use of vital dyes is still of great value in delineating macrophages either *in vivo* or *in vitro*. However, vital dyes may be taken up by other types of cell, such as fibroblasts, megakaryocytes, polymorphs and endothelial cells, although these do not have the same capacity for intracytoplasmic concentration as do the macrophages. The use of metallic impregnation techniques is limited to fixed tissues, since they tend to obscure cellular detail, but they are almost specific for macrophages and do not normally stain unrelated cells. It would appear that it is advantageous to combine vital staining and metallic impregnation techniques to obtain the maximum information by light microscopy (Marshall, 1956).

For morphological studies, it is more profitable to examine smears of macrophages from normal peritoneal or pleural fluid, or of macrophages which emigrate on to glass coverslips placed on an abraded area of skin using the 'skin window' technique (Rebuck and Crowley, 1955). In stained preparations, the nuclei of macrophages may vary from round to deeply indented structures with a condensed or vesicular chromatin pattern (Figs. 9, 10). The cytoplasm may be slightly basophilic or slightly eosinophilic, and often contains vacuoles and ingested material. Appro-

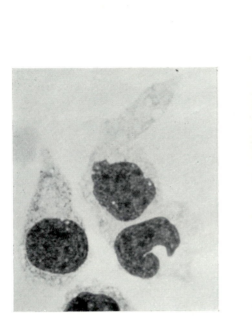
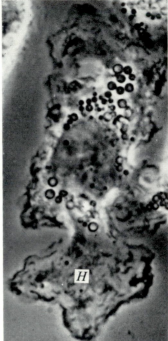

Fig. 10

Fig. 11

Fig. 10. Detail of cells shown in Fig. 9. Shows variable nuclear morphology and abundant vacuolated cytoplasm. Giemsa. × 2,200.

Fig. 11. Motile guinea-pig macrophage in culture. Shows hyaloplasmic veil (*H*) and numerous ruffles at the cell periphery. Phase contrast. × 3,000.

priate special stains reveal sparse mitochondria, the Golgi apparatus and prominent centrosome.

Macrophages are easily cultured *in vitro*. Methods of culture and macrophage behaviour *in vitro* have been reviewed by Jacoby (1965). They adhere tenaciously to glass and spread as monolayers, allowing good observation by phase contrast microscopy of living and fixed preparations. Phase contrast microscopy of the living cells reveals that the cell is motile, and that there is marked activity of the peripheral cytoplasm, which is often devoid of organelles, and exhibits slow undulations and ruffles which are continuous and move in a centripetal fashion. Single large fan-shaped projections of cytoplasm (hyaloplasmic veils) may extend from the edges of macrophages in motion (Fig. 11). Pseudopodia extend from the cell and may contain slim rod-like phase-dense mitochondria (Fig. 12). In the perinuclear region, mitochondria are

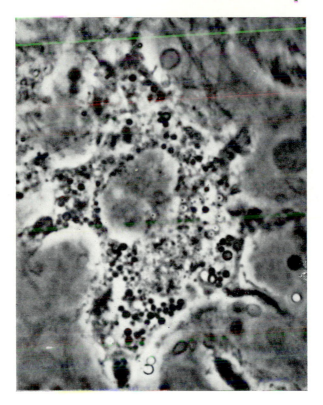

Fig. 12. Living sessile human macrophage in culture showing dendritic processes extending from cell periphery, abundant phase-dense organelles, and some phase-lucent vacuoles. Phase contrast. × 3,000.

intermingled with phase-dense rounded granules, probably lysosomes, and phase-lucent vacuoles. Lipid droplets are often present and large phago-somes are commonly observed. Phagocytic vacuoles containing whole cells or assorted particles are often seen.

The monocytes differ from the macrophages in that they are smaller cells and are less heterogeneous in morphology. They have less cytoplasm with smaller cell processes, and contain a smaller number of dense bodies and mitochondria.

ELECTRON MICROSCOPY

There is a large literature describing various aspects of macrophage ultrastructure. It would appear that certain features of the cell ultrastruc-ture are common to all macrophages (Fig. 13), but other features, such as the relative proportions of rough and smooth endoplasmic reticulum,

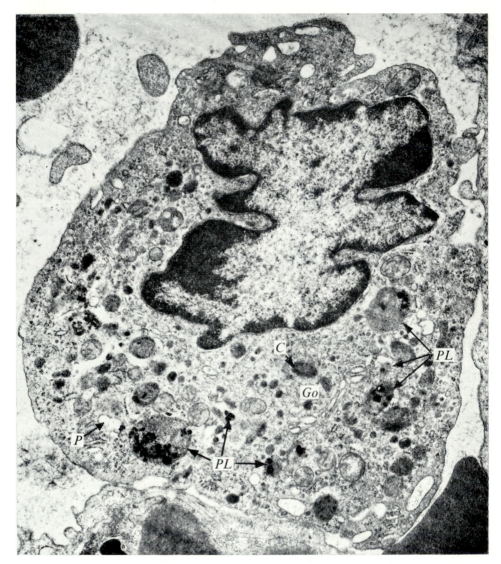

Fig. 13. Alveolar macrophage in lung of guinea-pig 24 hours after intravenous injection of carbon. Shows phagosomes (*P*) and phagolysosomes (*PL*) containing carbon particles, Golgi apparatus (*Go*), centrosome (*C*) and large numbers of granules varying in size and electron density. × 15,000.

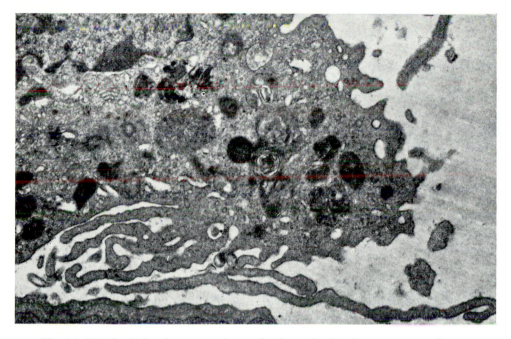

Fig. 14. Detail of alveolar macrophage of guinea-pig showing numerous cell processes. × 15,000.

may vary depending on the anatomical location of the cells and on physiological and pathological states. The appearances of cultured cells can vary according to the type of medium and duration of culture. Moreover, there is some evidence that there are minor species variations in macrophage ultrastructure.

The plasma membrane is about 80Å in thickness and comprises two electron-dense layers separated by a relatively clear layer. The plasma membrane is often very irregular and may be arranged as numerous microvilli and larger pseudopodia (Fig. 14), and there are often deep invaginations present. Study of the surface of peritoneal macrophages has been carried out with the scanning electron microscope (Carr, Clarke and Salsbury, 1969), and shows that the cells may vary widely in overall shape according to the conditions of stimulation. The cells are covered by ridgelike elevations which probably correspond to the ruffles seen during examination of living cells *in vitro*, and larger flangelike processes are also seen, which may correspond with the hyaloplasmic veils seen in macrophages moving *in vitro* (Fig. 15). The ridges and flangelike processes become more prominent when the cells are stimulated by lipids (Carr *et al.* 1969). The latter appearance was thought to represent

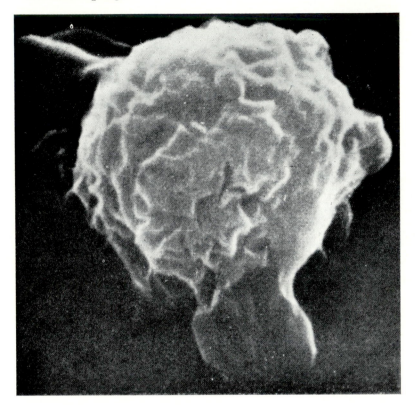

Fig. 15. Scanning electron micrograph of peritoneal macrophage showing peripheral flangelike processes and surface ridges. × 17,750. Reproduced with permission from Carr (1970).

an increase in the deformability of the cell which could be associated with an increase in its phagocytic ability.

On the outer surface of the plasma membrane is a prominent cell coat, best seen after ruthenium red staining (Carr, 1970) (Fig. 16), of amorphous material 8–160Å in thickness and with occasional areas of poorly defined fibrillary material. At least some of this material exhibits the staining characteristics of acid mucopolysaccharide, and this may be responsible for the marked adhesive properties of macrophages. Using the technique of electron microscope cytochemistry, North (1966a) demonstrated that the outer components of the plasma membrane of guinea-pig peritoneal macrophages exhibit adenosine triphosphatase (ATPase) activity. The enzyme was active in the presence of calcium and magnesium ions, which are known to play an important role in phagocytosis. ATPase activity was inhibited by treatment with sulphydryl poisons or trypsin,

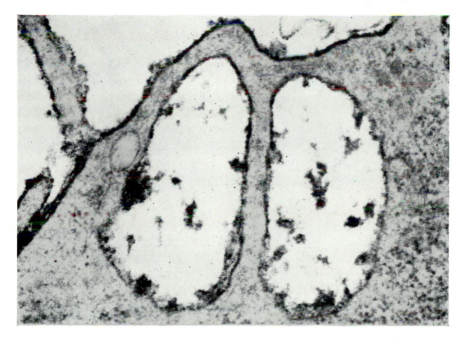

Fig. 16. Detail of macrophage stained with ruthenium red showing dense material adherent to outer layer of cell membrane. The coat is irregular in thickness, and there are also small fibrils present in a deep indentation. × 64,000. Reproduced with permission from Carr (1970).

and both of these factors are known to inhibit phagocytosis. Moreover, the adherence of phagocytic cells to glass is also inhibited by sulphydryl poisons. On the basis of these observations, North (1966a) suggested that cell membrane ATPase of macrophages may act during phagocytosis as a mechanoenzyme.

The indented nucleus exhibits peripheral condensation of chromatin and contains one or more nucleoli. It is surrounded by a membrane having numerous pores. The centrosome and Golgi apparatus are usually situated in the cytoplasm in the indentation of the nucleus, an area sometimes referred to as the 'hof' of the nucleus. The centrosome is rounded and exhibits fine granularity (Fig 13). The Golgi apparatus is variable in size and is composed of one or more stacks of parallel smooth membranes or lamellae, frequently with smooth membrane-bounded vesicles among the lamellae. Some of these vesicles appear to bud from the lamellae and may be related to primary lysosome formation (Cohn, Fedorko and Hirsch, 1966). The lysosomes are membrane-bound 'packets' containing enzymes. They are electron-dense and are bounded by a single unit membrane. The mitochondria are slim and sparse, but increase in size and number in cultured cells. There is a moderate amount of rough-surfaced endoplasmic reticulum (ER) with attached ribosomes. Some free ribosomes are present but polysomes are rarely observed. Little smooth-surfaced ER is present. Microtubules may be seen in the peripheral cytoplasm. The other major cytoplasmic component consists of electron-opaque granules of various sizes and densities. They are usually located in the region of the Golgi apparatus (Fig. 13). Some are homogeneous but others are multi-vesicular structures. There are frequently many small vesicles scattered throughout the cytoplasm. Some are electron-lucent and resemble pinocytotic vesicles whereas others contain dense material and may represent smaller variants of the larger dense granules.

Bundles of fine filaments, each filament 50–60 Å in diameter, are occasionally observed arranged in linear fashion in the perinuclear region. They may form a ring around the nucleus or Golgi apparatus, and may be closely approximated to mitochondria.

Macrophages from various sites frequently contain ingested material. In addition to ingested material, the endocytic vacuoles frequently contain lysosomal contents which have been discharged into the vacuoles, which may then be termed phagolysosomes or secondary lysosomes.

Since the peritoneal cavity is widely used as a source of macrophages for *in vivo* and *in vitro* studies, it is important to note that the majority of unstimulated peritoneal macrophages appear to be immature and resemble the blood monocytes. They contain a moderate amount of rough-surfaced ER, a small but well-defined Golgi apparatus and a few small electron-opaque granules in the cytoplasm. Macrophages cultured *in vitro*, and macrophages obtained from the peritoneal cavity following the intraperitoneal injection of various stimulants, show marked changes in ultrastructure. Most prominent is the formation of large electron-opaque granules, some of which have a complex matrix containing both electron-opaque and lucent vesicles. In addition, there is an increase in size of the Golgi apparatus with the appearance of new lamellae and tiny smooth-surfaced vesicles. Mitochondria increase in number. Large lipid droplets are found in apposition to the rough ER. The formation and size of electron-opaque granules *in vitro* is increased by raising the concentration of serum in the medium (Cohn, Hirsch and Fedorko, 1966). The cytoplasmic processes also become elongated and more numerous when the cell is stimulated.

Many, if not all, of the electron-dense granules observed in macrophages are thought to be secondary lysosomes and are probably derived from endocytic vacuoles. The multivesicular bodies and myelin figures which may be present may have two possible origins. The multivesicular bodies may arise when membrane-containing structures are engulfed by the cell in a heterophagic process during which extracellular material foreign to the macrophage is ingested; whereas myelin figures may result from a purely intracellular autophagic process (de Duve and Wattiaux, 1966) when a portion of the macrophage cytoplasm, including mitochondria, ER, or other organelles, is enveloped by a pinocytotic vesicle or lysosome, and is subsequently degraded within a membrane-bounded structure. In other cases it appears that small invaginations of the wall of a pinocytotic vacuole can occur, become pinched off, and result in multivesicular bodies. Thus it appears that both heterophagic and autophagic processes can lead to the same morphological end result.

METABOLISM OF MACROPHAGES

ENERGY METABOLISM

Studies on the sources of energy of macrophages have been generally carried out using suspensions of relatively pure populations of macrophages obtained by lavage of the peritoneal cavity or pulmonary alveoli. Macrophages from these two sources exhibit differences in their biochemical activities. In a comprehensive review of the metabolism of phagocytic cells, Karnovsky (1962) compared some of the metabolic activities of alveolar and peritoneal macrophages. He cited evidence which indicated that oxygen uptake was considerably higher in alveolar macrophages (30 μl/h mg protein) than in peritoneal macrophages (7.5 μl/h mg protein) when the cells were in a 'resting' state. Phagocytosis was suppressed in both cell types when glycolytic inhibitors were added to the system. However, a state of anaerobiosis, or the addition of cyanide or dinitrophenol, suppressed phagocytosis only in the alveolar macrophages, suggesting that alveolar macrophages depend on oxidative energy metabolism for phagocytic function. The oxygen consumption of peritoneal macrophages is depressed by the addition of glucose to the medium (Crabtree effect), whereas that of alveolar macrophages is increased (Pasteur effect) (Oren, Farnham, Saito, Milofsky and Karnovsky, 1963). Alveolar macrophages have a greater ability than peritoneal macrophages in converting glucose-1-^{14}C and glucose-6-^{14}C to ^{14}CO$_2$. The ratio of ^{14}CO$_2$ production from glucose-1-^{14}C to that from glucose-6-^{14}C is 6:1 for alveolar macrophages and 20:1 for peritoneal macrophages (Oren *et al.* 1963: Karnovsky and Wallach, 1963). In a histochemical study, Dannenberg and Walter (1961) demonstrated that alveolar macrophages in the rabbit possess generally larger quantities of aminopeptidase, cytochrome oxidase, succinic acid dehydrogenase and acid phosphatase than peritoneal macrophages. In an analysis of cells disrupted by freeze-thawing, Leake, Gonzalez-Ojeda and Myrvik (1964) showed that rabbit alveolar macrophages contained more lysozyme, acid phosphatase and beta-glucuronidase than peritoneal macrophages.

The above evidence supports the concept that the differences in metabolic behaviour between alveolar and peritoneal macrophages are relevant to the anatomical location of these cells and to their possible origin. The alveolar cells would have an adequate and continuous supply of oxygen, and are adapted to derive their energy by aerobic means mainly from oxidative phosphorylation. In contrast, peritoneal macrophages would have a comparatively limited supply of oxygen available and would obtain their energy mainly by anaerobic means from glycolysis. Consistent with these findings, alveolar macrophages possess greater numbers of mitochondria than do peritoneal macrophages.

It has not been established from these studies whether alveolar and peritoneal macrophages predominantly represent two morphologically and enzymatically distinct populations of cells, or whether macrophages can adapt themselves to function under aerobic or anaerobic conditions as circumstances demand. Until recent years, the view was generally held that alveolar macrophages originate in the lung itself, but there is now increasing evidence indicating that the alveolar macrophage population may be largely derived or augmented from macrophages and their precursors situated elsewhere in the body. In this connection Pinkett *et al.* (1966) produced evidence that alveolar macrophages can arise locally and from blood-borne antecedents. CBA mice were irradiated and injected intravenously with syngeneic (genetically identical) CBA bone marrow cells carrying the T6 chromosome marker (a translocated chromosome recognizable by the presence of a short fragment, but which causes no detectable functional abnormality). About two-thirds of the alveolar macrophages in mitosis carried the T6 marker and one-third did not. Since the dose of irradiation would have prevented the production of precursors by the recipient's own marrow, it was presumed that the cells not carrying the T6 marker arose locally. Moreover, the experiments of Russell and Roser (1966) have shown that intravenously injected alveolar and peritoneal macrophages are capable of settling in the recipient liver where they are morphologically and functionally identical with the indigenous Kupffer cells. From these findings it would appear that macrophages can derive energy facultatively from oxidative or glycolytic pathways as circumstances demand.

Relative to the numbers of metabolic studies on alveolar and peritoneal macrophages, few investigations have been carried out on the metabolic activities of monocytes and the macrophages of solid tissues because of the difficulties in obtaining large enough samples of pure populations of these cells. Bennett and Cohn (1966) have demonstrated active aerobic glycolysis in horse monocytes, but since the monocyte has little in the way of glycogen stores, it is largely dependent on exogenous substrates.

Cytochemical studies by various workers have demonstrated the presence of a large variety of enzymes within tissue macrophages, and have indicated considerable variations in the proportions of individual enzymes in macrophages at various sites under various conditions. The findings reflect considerable quantitative biochemical heterogeneity among macrophages which may reflect an adaptation to local environmental conditions rather than the existence of biochemically different populations of these cells.

LYSOSOMAL ENZYMES

Lysosomes are a heterogeneous group of cytoplasmic organelles which play a prominent part in the digestive and lytic processes of the cell, particularly in cells which take up material by endocytosis. Thus macrophages, which have a highly developed capacity for endocytosis, generally contain abundant lysosomes.

Biochemically, lysosomes have a rich content of acid hydrolases. Of these, acid phosphatase is the most commonly visualized by light and electron microscopy using histochemical methods by which it can be very precisely localized (Fig. 17), and the most commonly measured using biochemical techniques. Indeed, it has been suggested that any intracellular organelle, bounded by a single unit membrane, and staining positively for acid phosphatase, should be considered as belonging to the lysosome group. However, acid phosphatase is absent from guinea-pig and chicken blood monocytes but appears during their development into mature macrophages on culture *in vitro* (Weiss and Fawcett, 1953). This enzyme is present in the monocytes of other species. Moreover, it has been consistently demonstrated that levels of acid phosphatase are higher in alveolar than in peritoneal macrophages.

In addition to acid phosphatase, macrophage lysosomes have been shown to contain lysozyme, lipase, cathepsin, acid ribonuclease, acid deoxyribonuclease, neuranimidase, β-glucuronidase, hyaluronidase, aryl sulphatase and non-specific esterase. The lysosomes of macrophages have also been shown to contain non-enzymatic material. For example, lysosomes exhibit a positive staining reaction for phospholipid using acid haematin (Novikoff, 1961). Moreover, PAS-positive material, resistant to treatment with salivary diastase, can be demonstrated in macrophages in various sites, and is particularly abundant in a proportion of these cells in the thymus (Metcalf, 1966).

The numbers and appearance of the lysosomes vary according to the physiological and pathological influences on the cells undergoing examination. Since lysosomes and their enzymes play a major role in macrophage function, it would appear that an understanding of their formation and turnover is of great importance.

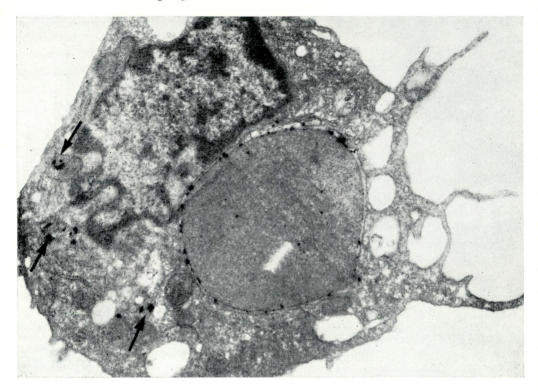

Fig. 17. Guinea-pig peritoneal macrophage which has ingested a sheep red cell in culture. Shows acid phosphatase reaction product around the margin of the vacuole containing the red cell, and also in cytoplasmic granules (arrows). × 7,500. Reproduced with permission from North (1966*b*).

An elegant study of this problem has been made by Cohn and his colleagues (reviewed by Cohn, 1968). They have shown that environmental factors to which cells are exposed played a major role in the formation of macrophage lysosomes. An important factor was the concentration of newborn calf serum in the medium. Cultures exposed to low levels of serum formed few dense granules and little or no enzymes, whereas high levels of serum (50 per cent) resulted in the rapid and extensive production of lysosomes and lysosomal hydrolyases. Examination of the cells exposed to high concentrations of serum by time-lapse cinematography revealed the origin of the dense granules. Shortly after explantation and exposure to the medium the cells began to pinocytose actively. Vesicles were seen to form at the cell membrane and streamed in unidirectional fashion into the perinuclear regions. Within a short time, large numbers of clear, phase-lucent pinocytotic vacuoles had accumulated about the Golgi complex. Here, they underwent a transition in density, shrinking somewhat in the process, and became typical dense granules. Whereas the initial lucent pinocytotic vacuoles were uniformly negative for acid phosphatase, the dense granules were strongly positive, suggesting the acquisition of the enzyme during the formation of the dense granules. It was therefore apparent that the dense granules or lysosomes arose from pinocytotic vacuoles.

When these investigators cultured macrophages in the presence of colloidal gold, there was extensive endocytosis, and the pinocytotic vesicles thus formed, containing particles of gold, streamed from the cell surface to the Golgi region. At this stage the pinocytotic vesicles contained no demonstrable acid phosphatase. The Golgi vesicles, distinguishable from the pinocytotic vesicles by their lack of contained gold, were seen to surround and apparently fuse with some of the pinocytotic vesicles. Acid phosphatase was demonstrable in the Golgi vesicles and saccules, and also within the dense bodies (secondary lysosomes or phagolysosomes) formed by the fusion of Golgi vesicles with pinocytotic vesicles. By means of electron microscopical auto-radiography, it was shown that new protein is synthesized in the rough ER, subsequently transferred ('shunted') to the Golgi apparatus, and eventually ends up as a dense granule. The presence of acid phosphatase activity in the Golgi complex was suggestive evidence that hydrolases follow a similar pathway.

The above data suggest that the primary lysosome of the macrophage is the tiny, smooth-surfaced, Golgi vesicle which is a package of enzyme which has not yet come into contact with substrate. The secondary lysosome contains both substrate and enzyme, and is composed of constituents of the extracellular environment in addition to the endogenously synthesized hydrolytic enzymes.

The nature of the stimulus to lysosome formation has been studied by Axline and Cohn (1970). The *in vitro* induction of lysosomal enzymes by phagocytosis was demonstrated in cultivated mouse peritoneal macrophages. Phagocytosis of sheep red blood cells and aggregated bovine gamma globulin produced a marked increase in the levels of acid phosphatase, β-glucuronidase and cathepsin D. Puromycin inhibited this enzyme response. In contrast, phagocytosis of polyvinyl toluene, polystyrene and insoluble starch particles produced no increase in macrophage lysosomal enzymes. The endocytic stimulus to lysosomal enzyme synthesis therefore occurred at or beyond the stage of digestion, and did not appear to be related to the fusion of secondary lysosomes with endocytic vacuoles. The observation that puromycin inhibits the enzyme increase after phagocytosis suggested that new enzyme synthesis was required, and is consistent with previous observations by Cohn *et al.* (1966*a*) that lysosome formation induced by pinocytosis is dependent upon new protein synthesis. The authors concluded that both the quantity of endocytosed material and its rate of enzymatic hydrolysis appear to control the level and persistence of lysosomal hydrolases.

The lysosomes may vary considerably in size and morphology. Secondary lysosomes (phagolysosomes) are large when the ingested material contained in the phagosomes is large. Fusion of secondary lysosomes may induce the formation of large, bizarre, complex bodies. Multivesicular bodies may possibly be formed in this way. The contents of secondary lysosomes may also condense after digestion to form bodies of smaller size.

Another structure belonging to the lysosome group is the autophagic vacuole. Such structures appear whenever a cell must sacrifice a portion of its own cytoplasm, and this may take place after a variety of stimuli. They are bounded by a single unit membrane and may contain other cytoplasmic structures such as mitochondria, Golgi vesicles and pieces of ER. They appear to acquire acid phosphatase as soon as the membrane surrounding them is formed.

Lysosomes (either secondary lysosomes or autophagic vacuoles) which have carried out digestive functions finally appear filled with debris, usually of a lipid nature. They are then termed 'residual bodies'. When residual bodies, which have been shown to be capable of taking part in further acts of endocytosis and digestion, amass several concentric membranous arrays which resemble the ultrastructure of myelinated nervous tissue, the final forms are referred to as 'myelin figures'. Extrusion of digestive debris from residual bodies to the cell exterior is termed 'exoplasmosis' or 'reverse pinocytosis'.

The continued presence of secondary lysosomes within macrophages is dependent on continued endocytic activity. When macrophages are placed in an environment where pinocytotic activity is suppressed, cells which have previously accumulated secondary lysosomes lose these organelles. The loss of dense granules and concomitant reduction in enzyme levels does not take place by extrusion into the surrounding medium (Cohn, 1968). It appears that, when the uptake of substrate is stopped, the organelle shrinks in size by means of intralysosomal digestion. Moreover, the lysosomal enzymes disappear as well. This reversibility of lysosomal formation and enzymatic activity within macrophages which *in vitro* depends on the degree of endocytic activity, can be postulated as an adaptation process required *in vivo* when 'resting' macrophages are faced with a sudden demand for increased hydrolase formation accompanying the necessity for increased phagocytic activity as, for example, during bacterial invasion of the tissues.

DEOXYRIBONUCLEIC ACID (DNA)

Mitosis in macrophages and factors which stimulate its increase are considered in Chapter 2. The use of tritiated thymidine has indicated that, under normal conditions, macrophages and monocytes do not incorporate appreciable amounts of thymidine. It has been reported by various workers that, in the absence of exogenous stimulation, between 1 and 4 per cent of peritoneal macrophages, monocytes and Kupffer cells will incorporate tritiated thymidine *in vivo* during DNA synthesis prior to cell division. *In vitro* studies have shown that peritoneal macrophages will incorporate thymidine in culture after a lag period of about 48 hours.

RIBONUCLEIC ACID (RNA)

The metabolism of RNA in macrophages has not been extensively studied. However, it has been established that unstimulated and stimulated peritoneal macrophages will incorporate labelled precursors into their ribonucleic acids. It has been shown that rabbit peritoneal macrophages can exhibit a rapid turnover of RNA, but at the same time may show no synthesis or turnover of DNA (Watts and Harris, 1959).

OTHER METABOLIC ACTIVITIES

The participation of macrophages in the metabolism of iron, proteins, lipids, sterols and steroids is considered in Chapter 5.

2

LIFE HISTORY OF MACROPHAGES

ORIGIN OF MACROPHAGES

In considering the origin of macrophages, two problems arise for the investigator. First, there is the problem of identifying the immediate antecedents of the macrophages which appear in various organs and in inflammatory exudates; and, secondly, there is the problem of identifying the ultimate precursor or primitive stem cell from which (presumably) all macrophages arise. Both problems are the subject of active investigation at the present time. Moreover, while much is now known about the origin and turnover of monocytes and macrophages which appear in response to inflammatory stimuli, relatively little is known about the origin and turnover of macrophages under normal conditions and in solid tissues.

There is now considerable evidence that the precursors of macrophages are principally located in the bone marrow. Balner (1963) implanted allogeneic (genetically different) bone marrow into irradiated recipient mice and showed that the peritoneal macrophages of the resultant chimeras (animals in which the haemopoietic tissues are partly or wholly of non-host genetic type) were of donor origin. In similar experiments, Goodman (1964) implanted bone marrow, foetal liver, peripheral blood leucocytes and peritoneal fluid cells from donor mice, and bone marrow from rats, into irradiated mice, and showed that in all chimeras which retained donor haemopoietic grafts, the peritoneal fluid cells were all of donor type.

Volkman and Gowans (1965a) employed the 'skin window' technique of Rebuck and Crowley (1955) to investigate the origin of macrophages in non-bacterial acute inflammatory exudates in rats. In this technique, a glass coverslip is applied to an abraded area of skin or into a pocket fashioned in the subcutaneous tissues. Polymorphs and macrophages migrate into the area and adhere to the glass coverslip which can then be removed for examination of the cells at any desired interval after the application of the coverslip. Using tritiated thymidine, Volkman and Gowans (1965a) showed that, 24 hours after applying the coverslip, up to 57 per cent of macrophages could be labelled by giving a single intravenous dose of tritiated thymidine, providing the injection was given

[32]

about one day before the application of the coverslip. This indicated that inflammatory macrophages originate from rapidly dividing precursors which are proliferating continuously. Experiments with rats in parabiosis showed unequivocally that almost all of the exudate macrophages were derived from the blood stream and did not arise by the multiplication of locally situated cells in the tissues. Further studies with tritiated thymidine showed that very few macrophages on the coverslips or macrophage antecedents in the blood were synthesizing DNA. These findings showed that the precursors of inflammatory macrophages proliferate at sites other than the area of inflammation and then release their progeny into the circulation.

Further studies by Volkman and Gowans (1965b) showed that lymphocyte depletion by chronic drainage from the thoracic duct or 400 rads of X-irradiation failed to suppress the emigration of macrophages or reduce the proportion of them which became labelled after an injection of tritiated thymidine. X-Irradiation with 750 rads suppressed the emigration and the labelling of the exudate macrophages. Both were restored to normal when the tibial marrow was shielded during irradiation. Radioactively-labelled cell suspensions obtained from thoracic duct lymph, lymph nodes, thymus, spleen and bone marrow were transfused into syngeneic recipients. The emigration of labelled macrophages on to coverslips could only be demonstrated in recipients of labelled bone marrow and spleen cells. Labelled monocytes were found in the blood of rats which had received injections of labelled bone marrow. From these findings, Volkman and Gowans concluded that, in the rat, the bone marrow, and to a lesser extent the spleen, are the major sources of macrophages which emigrate into foci of acute non-bacterial inflammation. From their findings they also concluded that the great majority of small lymphocytes, which are long-lived recirculating cells, cannot be the antecedents of macrophages.

Volkman (1966) investigated tritiated thymidine labelling in individual and parabiotic rats and showed that the macrophages in peritoneal exudates were derived from cells in the blood which were the progeny of rapidly and continuously proliferating precursors. Cells in the exudates which were morphologically indistinguishable from small lymphocytes were also found to have the labelling features of a rapidly proliferating population, in contrast with the known kinetics of the majority of small lymphocytes in blood and thoracic duct lymph. The experimental evidence indicated that the lymphocyte-like cells had emigrated from the blood and that bone marrow was a source of their precursors. Virolainen (1968) also concluded that mouse peritoneal macrophages are derived from the bone marrow. He injected bone marrow cells into lethally-irradiated recipients and observed that the donor cells, identified by their distinctive T6 chromosome, gave rise to peritoneal macrophages. Lymph node cells failed to provide a source of macrophages.

Van Furth and Cohn (1968) found that a peak of labelled mononuclear phagocytes occurred in the bone marrow 24 hours after a single pulse of tritiated thymidine. This was followed 24 hours later by a peak of labelled monocytes in the peripheral blood. They concluded that rapidly dividing mononuclear phagocytes of the bone marrow, the promonocytes, are the progenitor cells of monocytes. A rapid entry of monocytes from the blood into the peritoneal cavity was observed after the induction of sterile inflammation, and they concluded that the monocytes transformed to peritoneal macrophages under these conditions. There was no indication that mononuclear phagocytes originate from lymphocytes. Further studies by Van Furth (1970) showed that, originating from the single division over 19 hours of large promonocytes in the bone marrow, daughter monocytes circulated in the blood with a mean transit time of 32 hours before populating the peritoneum, lungs and liver, where they could persist for 40–60 days. Therapeutic doses of corticosteroids inhibited the migration of monocytes from the marrow via the blood and their subsequent influx into inflamed areas, but had no effect on the proliferation of promonocytes which continued to divide in the bone marrow although their offspring did not appear in the peripheral circulation. He also demonstrated that some of the small lymphocyte-like cells of the peripheral blood are in fact macrophages. Further studies by Van Furth and his colleagues (Van Furth, Hirsch and Fedorko, 1970; Van Furth and Diesselhoff-Den Dulk, 1970) have shown that, in the mouse, the promonocytes of the bone marrow are proliferating cells, whereas under normal conditions the monocytes of bone marrow and blood are non-dividing cells. They concluded that promonocytes are multiplicative cells (i.e. divide to give rise to morphologically different cells), and that the division of one promonocyte gives rise to two monocytes. In all probability the promonocytes divide only once. The pool of promonocytes appears to be maintained by the influx of cells from a pool of precursor cells, the identity of which had not been established since, unlike promonocytes and monocytes, they do not stick to glass. It was also found that the size of the monocyte pool of the bone marrow is about two times greater than the peripheral blood monocyte pool. The generation time of promonocytes is about 19.5 hours, and the calculated turnover time of the monocytes in the bone marrow compartment is about 55.2 hours. However, the monocytes do not leave the bone marrow in regular fashion as they are produced, and it appears that one fraction of bone marrow monocytes is fed into the peripheral blood rapidly, whereas another fraction arrives after some delay. The promonocytes have a diameter of 14–$20\,\mu$. They engage in pinocytosis and phagocytosis, but are less active in these functions than are peripheral blood monocytes or peritoneal macrophages. Promonocytes also possess peroxidase-positive granules; blood monocytes possess a smaller number of peroxidase-positive granules, and various types of mature macrophages are peroxidase-negative.

Vernon-Roberts (1969 a) showed that, after a single subcutaneous injection of oestrogen in mice, two peaks of increased numbers of small lymphocytes were observed in the peritoneal cavity on the first and sixth days after oestrogen injection. The peaks of increased numbers of small lymphocytes were followed 24 hours later by an increase in the number of cells intermediate in appearance between lymphocyte and macrophage, and 48 hours later by an increase in the number of mature macrophages. Both peaks of increased lymphocytes were accompanied by an increase in the percentage of small lymphocytes which had incorporated tritiated thymidine in the peritoneal cavity and in the blood stream. No changes were observed in the labelling of blood monocytes and large lymphocytes. The findings suggested that oestrogen had stimulated the proliferation of two populations of small lymphocytes which had migrated by means of the blood stream to the peritoneal cavity where they transformed to macrophages. In the presence of viable neutrophil polymorphonuclear leucocytes, lymphocytes obtained from the peritoneal cavity were shown to transform to macrophages *in vitro*. To aid the identification of phagocytic cells, a suspension of colloidal carbon was added to some of the cultures 2 hours before killing. Not all the lymphocyte-like cells underwent transformation, and the maximum yield of macrophages was 39 per cent in 48-hour cultures. The findings were taken to support the lymphocytic origin of some of the peritoneal macrophages, and suggested an important role for the relatively few polymorphs seen in the non-inflamed peritoneal cavity. In the same study, the mobilization of peritoneal macrophages to distant inflammatory sites was investigated by injecting carbon-labelled donor macrophages into recipients in whom 'skin-window' coverslips were inserted subcutaneously. Examination of the coverslips at 24 and 48 hours revealed that many carbon-containing donor macrophages had been mobilized from the peritoneal cavity and were adherent to the coverslips. From this series of experiments, it was concluded that cells indistinguishable from small lymphocytes migrated by the blood stream to the peritoneal cavity where they transformed to macrophages with the aid of an essential factor provided by polymorphs. The finding also suggested that mature peritoneal macrophages may be mobilized to local or distant inflammatory sites when required; and, if this necessitates migration through the blood, it would not be unreasonable to assume that the circulating macrophages would be indistinguishable from blood monocytes.

If extra-vascular cell transformation in the peritoneal cavity plays an intermediate part in the transformation of bone marrow lymphocytes to blood monocytes, the above findings would reconcile the finding that the bone marrow is the major site for the production of inflammatory macrophage precursors (Volkman and Gowans, 1965*a, b*; Volkman, 1966) with the known paucity of identifiable monocytes in normal marrow, and the fact that the macrophages of early inflammatory exudates are derived from monocyte-type cells as opposed to small lymphocytes (Volkman and Gowans, 1965*a, b*; Spector and Coote, 1965). It would also explain the delay in emergence of newly formed monocytes in the blood in response to inflammatory stimuli (Volkman and Gowans, 1965*a*).

The above findings, taken as a whole, also appear to establish beyond doubt that the blood-borne cells which are the immediate antecedents of macrophages in inflammatory exudates are most likely to be monocytes. The view that blood-borne lymphocytes transform to exudate macrophages has previously been put forward by many workers and there is no doubt that cells indistinguishable from small lymphocytes are seen in acute and chronic inflammatory lesions. However, the evidence on lymphocyte to macrophage transformation in inflammatory exudates has been largely deduced from serial morphological observations of the apparent simple progression of cells intermediate in appearance between small lymphocytes and macrophages in experimental inflammatory lesions. The labelling data in the experiments of Volkman and Gowans (1965*a*) quoted above are not consistent with this view, since they found that not more than 2 per cent of small lymphocytes in the blood were labelled at any one time, whereas over 50 per cent of the exudate macrophages were labelled. The patterns of labelling of blood monocytes and exudate macrophages were very similar, and differed markedly from that of small and large lymphocytes in the blood. Thus it is quite clear that evolution into macrophages is not a property of the great majority of lymphocytes.

The work of Spector and his colleagues (reviewed by Spector, 1969) has provided additional evidence that blood monocytes are the immediate antecedents of inflammatory macrophages. In inflammatory reactions to various forms of intradermally injected irritants, they found that the majority of macrophages were derived from blood monocytes which had previously been labelled by the injection of large amounts of colloidal carbon. Experiments with tritiated thymidine by these workers were also consistent with this conclusion. In chronic inflammatory lesions of some standing, the findings are complicated by the occurrence of mitosis within the exudates. The fact that Volkman and Gowans (1965*a*) showed that less than 0.5 per cent of macrophages which had already migrated on coverslips could take up tritiated thymidine, tells against the possibility of local macrophage proliferation making a significant contribution to exudate populations in the case of acute inflammatory lesions; however, increasing evidence is accruing which suggests that mitosis in macrophages in long-standing inflammatory lesions may contribute considerably to the new macrophages which appear in such areas.

From the above studies it may be deduced that the precursor cell for the majority of macrophages which appear in early inflammatory exudates is situated in the bone marrow, although from these studies it is also clear that some macrophage precursors could be sited elsewhere, for example in foetal liver and peritoneal fluid cells (Goodman, 1964), or in the spleen (Volkman and Gowans, 1965b). However, the findings of Goodman (1964) indicate that those precursors which are present in extramedullary tissues may be considered as 'resting wandering cells' (Maximow, 1962) which are probably multipotential cells which have themselves migrated into the tissues from the bone marrow.

The morphological identity of the ultimate macrophage precursor cell in the bone marrow is unknown. The bone marrow contains cells having a wide variety of morphological appearances, and cells indistinguishable from small and medium lymphocytes are present in considerable numbers, as are cells of monocytoid appearance. Another possible candidate for the macrophage stem cell is the reticulum cell (probably synonymous with the 'resting wandering cell' of Maximow) which almost certainly has pluripotential capabilities. The great majority of lymphocyte-like cells in the bone marrow are apparently formed therein (Osmond and Everett, 1964), and their exact function is unknown. The experiments of Volkman (1966) demonstrated that infused bone marrow was the source of significant numbers of lymphocyte-like cells in induced peritoneal exudates, contrasting with the largely negative yields from infusions of thoracic duct cells. In this connection, a stem cell function has been attributed to a special group of small lymphocytes present in the bone marrow (Cudkowicz, Upton, Shearer and Hughes, 1964) which exhibit the fast turnover which qualifies them as being possible precursors of macrophages. It is not known whether this fast turnover group of lymphocytes is homogeneous nor what proportion of these cells is capable of response to acute inflammation. The observations of Vernon-Roberts (1969a) that lymphocyte to macrophage transformation takes place in the peritoneal cavity and may precede the mobilization of peritoneal macrophages to inflamed areas, provides one possible explanation that the precursor cell in the bone marrow has the morphology of a small lymphocyte, whereas the cells migrating from the blood into areas of inflammation have the morphology of monocytes. In this connection the majority of macrophages in the peritoneal cavity have been shown to have the general morphology and cytochemistry of the monocyte as opposed to the mature macrophage (Cohn, 1968). Forbes (1965) showed that when sensitized mice are challenged with antigen the peritoneal macrophages divide at a rapid rate and incorporate tritiated thymidine. His evidence also suggested that some new macrophages were derived by the transformation of small lymphocytes. Thus there is circumstantial evidence that macrophages in acute inflammation induced by mild trauma to the skin and those evoked during immunological responses are derived from identical precursors.

The evidence quoted above overwhelmingly favours the bone marrow as being the site of origin of the great majority of macrophages. However, there is considerable evidence that macrophages initiate some primary immune responses by taking up antigens, and that this may be an essential prelude to antibody synthesis by other cells; but studies on the functions and characteristics of thymic cells suggest that cells which react initially with antigens originate in the thymus and are not directly derived from the bone marrow (Miller and Mitchell, 1968; Mitchell and Miller, 1968). Thus, if macrophages are the cells which initiate immune responses, then one can postulate an additional population of macrophages derived from bone marrow cells which are 'conditioned' by the thymus. In this latter connection, Miller and Mitchell hold the view that there is a steady stream of cells from the bone marrow to the thymus and to secondary lymphoid tissues such as the lymph nodes and spleen. If the above postulates are correct, it would appear that an obvious candidate for a thymic-derived macrophage is the 'antigen-retaining reticular cell' of lymph node follicles. These cells are situated in close approximation to lymphocytes, and appear to play a major role in taking up and retaining antigen in lymph nodes prior to antibody synthesis (Nossal, Abbot and Mitchell, 1968; Nossal, Abbot, Mitchell and Lummus, 1968). It is also of relevant interest that, in studies on the derivation of macrophages in inflammation due to Freund's complete adjuvant (water-in-oil emulsion with added *M. tuberculosis* or related organism), Spector and Willoughby (1968) demonstrated that in the earlier phase of the reaction all the macrophages were found to be marrow-derived, and no evidence was obtained that lymph node cells entered the lesion. However, in lesions of over 4 weeks' duration small numbers of lymph node cells entered the lesion from the blood stream, but thymic cells could not be demonstrated to behave in this way. In contrast, Freund's incomplete adjuvant (water-in-oil emulsion without bacillary addition), although provoking a normal influx of bone marrow-derived cells, did not stimulate the same influx of lymph node-derived cells in the long-standing granuloma as was observed with the complete adjuvant. These findings indicate differences in the cell populations of chronic inflammatory lesions produced by non-antigenic stimuli (incomplete adjuvant) and chronic inflammation complicated by the presence of antigen (complete adjuvant), and suggests the possibility that antigen may induce the additional emigration of lymph node cells (possibly derived originally from the thymus) in order to process or otherwise deal with the antigen.

There is evidence that hepatic and splenic macrophages may themselves proliferate under abnormal conditions. It has been demonstrated that fixed macrophages in the liver of the mouse undergo mitosis in large numbers during an infection with *Listeria monocytogenes*, and that the macrophage mitotic response always precedes the manifestation of efficient host immunity against the infection (North, 1969*a*, *b*). It is well recognized that both hepatomegaly and splenomegaly are a marked feature of the pathology of graft-versus-host reactions, and an overall increase in the number of macrophages in these organs is apparent. It has been shown that mice with graft-versus-host reactions induced by injecting thoracic duct cells exhibit increased mitosis in liver macrophages, and the dividing cells are predominantly of donor origin (Howard, Christie, Boak and Evans-Anfom, 1965). In contrast, Fox (1962) reported that, in the spleens of mice with graft-versus-host reactions induced by foreign spleen cells, the proliferating cells were largely of host origin.

It is clear that a great number of uncertainties exist regarding the origin of macrophages in various tissues under normal and pathological circumstances, not only in regard to the identity of macrophage precursors but also to their sites of origin.

LIFESPAN AND TURNOVER OF MONOCYTES

It has been found that, in rats and mice, originating from the single division over 19 hours of large promonocytes in the bone marrow, daughter monocytes circulate in the blood with a mean transit time of 32 hours before populating the peritoneum, lungs and liver (Van Furth, 1970). The monocytes leave the circulation in random fashion, with a half time of 22 hours (Van Furth and Diesselhoff-Den Dulk, 1970).

Whitelaw (1966) injected rats repeatedly with tritiated thymidine and, after a period of 8 days, found that the majority of circulating monocytes contained tritiated thymidine. The injection of thymidine was then stopped and the rate of loss of labelled and unlabelled cells was examined over the next 10 days. The results showed that approximately 3.6×10^6 monocytes were produced each day, and that the half-life of the blood monocyte is in the order of 3 days. Additional information obtained from these studies indicated that the new monocytes are generated over a period of about 24 hours, are derived from precursors which divide about three times before the circulating blood monocyte is formed, and that the cells leave the circulation in random fashion. The lifespan of 3 days for the blood monocyte quoted above agrees with the findings of Volkman and Gowans (1965*a*) following single injections of tritiated thymidine.

There is limited evidence that recirculation of monocytes takes place, so that cells which have previously left the circulation may return to the blood stream. It appears quite certain that this process does not occur to the same extent as the recirculation of small lymphocytes.

TRANSFORMATION OF MONOCYTES TO MACROPHAGES

In the section dealing with the origin of macrophages, the evidence was quoted which favours the monocyte as being the immediate precursor of inflammatory macrophages. The fate of the monocyte in the absence of inflammation is unknown. Apart from a limited recirculation of these cells, it would not be unreasonable to suggest that a proportion of them normally emigrate into the tissues and become transformed to, and indistinguishable from, mature macrophages. It is also possible that some monocytes attach themselves to the walls of blood sinusoids in organs such as the liver, spleen and bone marrow, and become indistinguishable from the macrophages normally resident in those sites. Apart from the recent isotopic studies, microscopic observations on the transformation of monocytes to macrophages *in vivo* were reported by Ebert and Florey (1939), and their findings have been confirmed by many others utilizing a variety of experimental inflammatory lesions and have more recently been complemented by studies on ultrastructural and enzymatic changes in inflammatory exudates. There have also been extensive studies of the maturation of monocytes to macrophages *in vitro*.

Early work on the development of tissue culture techniques revealed that the white cells of peripheral blood served as a readily available source of monocytes which could be cultured. Cells obtained from buffy coat (the thin pale layer of white cells situated at the junction of plasma and erythrocytes after centrifugation of whole blood) are readily maintained in culture for many days. During the early days of these cultures, the monocytes readily adhere to glass and become easily separable from the other leucocytes and platelets which disintegrate or are washed away during medium changing. The resulting relatively pure population of monocytes actively phagocytose cell debris and accumulate increasing numbers of dense bodies within their cytoplasm. There is also an increase in the amount of cytoplasm with progressive enlargement of the cells, until the cell population becomes indistinguishable from mature macrophages. This process occurs without evidence of cell division occurring in the monocytes. However, it is of interest to note that multinucleate giant cell formation, resulting from the fusion of cells is not an uncommon feature in cultures of monocytes and macrophages, and is particularly prone to occur when cultures are prolonged.

A more detailed study of monocyte changes in culture was made by Weiss and Fawcett (1953) who studied the changes in these cells obtained from buffy coats of chicken blood. They observed an increase in the number of lipid-containing PAS (periodic acid–Schiff)-positive inclusions, and the appearance of acid phosphatase, which had previously been absent, as the cells matured from monocytes to macrophages. Sutton and Weiss (1966) carried out an electron microscopic study of the same type of culture which they examined at frequent intervals as the culture was maintained over a period of 70 days. They described the transformation of monocytes into macrophages, and then the transformation of the latter into epithelioid-type macrophages and multinucleate giant cells. Changes were observed in the Golgi region which appeared to have been associated with an increase in the number of lysosomes as the cells matured from monocyte to macrophage, but the number of lysosomes diminished as the macrophages progressively transformed to epithelioid and giant cell forms. An increase in the number of mitochondria was also observed.

A combined morphological and biochemical study on horse blood monocytes was performed by Bennett and Cohn (1966). Within 24 hours of commencing culture, the monocytes exhibited increased numbers of acid phosphatase-positive perinuclear granules which had the properties of lysosomes, and this was accompanied by a quantitative increase in the accumulation of lysosomal enzymes. Mitochondria also increased in number, and this was accompanied by a twofold increase in the activity of the mitochondrial enzyme cytochrome oxidase. These morphological and chemical changes were accompanied by an increased capacity of the cells to ingest bacteria and particles of colloidal gold. Similar *in vitro* studies were carried out by Cohn and his colleagues (reviewed by Cohn, 1968) on the 'mononuclear phagocytes' obtained from the peritoneal cavities of mice. These cells have many of the morphological attributes of the blood monocyte in terms of size, nuclear morphology and cytoplasmic organelles, but differ in that they generally contain a few more dense granules. Within 24 hours of commencing culture, mitochondria increased in length and there was an increase in the number of lysosomes in the perinuclear region. As culture proceeded, there was a progressive increase in the size of the cell and a further increase in the number of lysosomes. Ultrastructural observations revealed an increase in the components of the Golgi complex related to the increased formation of lysosomes. A fifty-fold increase in the amount of acid phosphatase occurred within three days of culture, and there was also an increase in the cell content of β-glucuronidase, cathepsin and protein. The cells also exhibited an increased avidity for the uptake of neutral red.

Conclusive or suggestive evidence of the transformation of monocytes to cells other than macrophages is lacking. Morphological and cytochemical evidence supports the view that the blood monocytes should be regarded as immature macrophages, and are not macrophage precursors in the sense that the maturation of monocyte to macrophage does not necessitate cell division, although it has been reported that monocytes actively divide in inflamed areas (Spector, 1969). Whether or not all monocytes mature to macrophages *in vivo* is unknown, but tissue culture studies, which must be interpreted with caution, suggest that the great majority of monocytes possess the capacity for transformation. The division between monocyte and macrophage is largely arbitrary and depends on the personal criteria of the observer.

LIFESPAN AND TURNOVER OF MACROPHAGES

The majority of mature macrophages, as opposed to monocytes which may be considered to be immature macrophages, appear to be long-lived cells. There are species differences in the ease with which macrophages can survive in tissue culture. Thus, rat and mouse macrophages may be kept alive in culture for a long period of time (at least 100 days) without cell division occurring on a large scale, but guinea-pig macrophages apparently do not survive for this period of time, either because they demand more critical environmental circumstances or because they are less long-lived than the macrophages of other species. The observations on pigment-laden macrophages in the tissues of experimental animals and man suggest that the individual macrophages may persist unchanged for a very long time, and possibly for many years. Moreover, the low rate of uptake of tritiated thymidine by Kupffer cells and peritoneal macrophages (less than 1 per cent labelled in 24 hours) indicates that these cells are normally replaced very slowly and suggests that they are long-lived cells.

Spector and his colleagues have made exhaustive studies of the turnover and proliferation of macrophages in inflammatory exudates (reviewed by Spector, 1969). Examination of the cellular changes occurring in adjuvant-induced granulomata was undertaken after the emigrating mononuclear cells had taken up tritiated thymidine prior to their entry into the lesions. The percentage of labelled mononuclears in the inflamed area remained constant for a few days and then fell in exponential fashion between the seventh and twenty-eighth day after induction of inflammation. Further studies indicated that the loss of labelled cells was due to cell turnover with replacement by mitotic division, as evidenced by a decline in the average nuclear grain count of the tritiated cells which was in parallel to the observed decline in the proportion of labelled cells. In inflammatory lesions of over 24 hours' duration, about 2 per cent of the macrophages incorporated tritiated thymidine, and thymidine incorporation was observed in macrophages, epithelioid cells and giant cells in lesions of up to 3 months' duration. In 48-hour lesions, the uptake of thymidine was predominantly in perivascular macrophages, but in older lesions the label was taken up predominantly by macrophages within the bulk of the granuloma. The findings suggested that, in inflammatory granulomata, there may be two phases of macrophage proliferation, one which takes place in newly-emigrated cells situated in the perivascular areas, and the other which occurs after the macrophages become established within the matrix of the lesions.

Studies on chronic inflammatory lesions induced by a variety of stimuli have suggested that the turnover of macrophages within the lesions is to some extent dependent upon the nature of the stimulus itself. Spector (1969) has reviewed the evidence for the existence of 'high turnover' and 'low turnover' granulomata, and has concluded that although low and high turnover reactions exist in distinct forms in mature lesions, there is a strong tendency for high turnover granulomata to evolve with time into low turnover lesions. It would appear that the smaller the dose of irritant particle, the sooner does this transformation occur. There appears to be a progressive selective sequestration of the irritant in a particular section of the granuloma population, and the progressive disappearance of cells containing little demonstrable irritant. There is a progressive increase in the preponderance of long-lived macrophages containing much endocytosed irritant; the evidence for the heavily-laden macrophages being long-lived was based on the co-existence of demonstrable RNA synthesis, absent DNA synthesis, low renewal rate via emigration, and long half-life as suggested by the rate of disappearance of previously labelled cells. These findings suggested the possible existence of two populations of macrophages; one with a high turnover rate liable to be damaged by endocytosed foreign material, and the other with a low turnover rate, a capacity to survive for long periods, and the ability to phagocytose very large amounts of material.

MITOSIS IN MACROPHAGES

The use of tritiated thymidine has indicated that, under normal conditions, macrophages do not incorporate appreciable amounts of thymidine, thus indicating that they are long-lived cells. It has been reported by various workers that, in the absence of exogenous stimulation, between 1 and 4 per cent of mouse peritoneal macrophages will incorporate tritiated thymidine during DNA synthesis prior to cell division *in vivo* and *in vitro*. Increased thymidine uptake and high mitotic rates may be induced by various stimuli, such as mechanical disturbance of the peritoneal cavity, the monocytogenic extract from *Listeria monocytogenes*, the subcutaneous injection of oestrogen, the subcutaneous injection of antigen in previously sensitized animals, and the intraperitoneal injection of glycogen or endo-toxin. Unstimulated Kupffer cells also exhibit a thymidine incorporation rate of 1–4 per cent, but these cells have also been shown to exhibit vigorous proliferative activity, evidenced by increased DNA synthesis and mitotic activity, after oestrogen injection (Kelly, Brown and Dobson, 1962) or *Listeria* infection (North, 1969 a, b).

It is significant that macrophages may exhibit marked mitotic activity when immunological mechanisms are involved. Forbes (1965) showed that when mice which had been sensitized with normal rabbit serum, ovalbumin or bovine serum albumin in Freund's adjuvant were challenged with a subcutaneous injection of the appropriate antigen, an increased rate of DNA synthesis in peritoneal macrophages and lymphocytes was apparent within 2 hours. Most of the cells synthesizing DNA were macrophages. Between 20 and 30 hours after the second injection of antigen, the number of macrophages which had incorporated tritiated thymidine had risen from a resting level of less than 2 per cent to a maximum of about 50 per cent. Similarly, macrophages obtained from mice immunized with bovine serum albumin exhibit increased DNA synthesis when cultured in the presence of the antigen (Rowley and Leuchtenberger, 1964). The peritoneal macrophages of mice previously exposed to infection with *Listeria monocytogenes* or *Brucella abortus* also undergo extensive mitosis when re-infected with the same organisms (Khoo and Mackaness, 1964). Even in the absence of secondary antigenic challenge, there is proliferation of peritoneal macrophages during primary exposure of mice to *Listeria monocytogenes* or *Bacillus Calmette-Guérin* (BCG), which are short-lived and long-lived infections respectively, and the newly-proliferated population of macrophages have an increased capacity to phagocytose inert particles and to spread on a foreign surface (North, 1969 a). In these latter experiments, it was apparent that the rate of initiation of cellular proliferation and production of activated macro-phages are determined by the metabolic characteristics of the infecting organisms. In chronic inflammatory lesions produced in rats by the subcutaneous injection of suspensions of killed tubercle bacilli, macro-phages exhibit increased uptake of tritiated thymidine which may be stimulated by antigenic constituents of the tubercle bacilli (Spector and Lykke, 1966).

There have been conflicting reports on the incidence of mitosis when macrophages are cultured. Macrophages obtained from various sources generally exhibit about 1 to 3 per cent of cells in mitosis, the number of mitoses tending to increase as the duration of culture increases. If the cells for culture are obtained from oil-induced peritoneal or alveolar exudates, then the incidence of cells in mitosis may increase markedly above 3 per cent during the first week of culture.

The overall findings on the occurrence of mitosis in macrophages indicate that these cells are generally long-lived slowly-replicating cells, but can undergo vigorous proliferation as a result of various specific (immunological) and non-specific stimuli. This capacity for rapid proliferation would seem to be a necessary part of the defensive reaction to challenge by micro-organisms and antigens.

CIRCULATION, MIGRATION AND
FATE OF MACROPHAGES

Under experimental conditions, it has been demonstrated that macrophages from one site are capable of migration to another, but whether this is a normal event is unknown. Careful examination of the peripheral blood in animals and man reveals the presence of occasional monocytoid cells which are larger and contain more mitochondria and lysosomes than the average blood monocyte, but, taking into account the arbitrary division between monocyte and mature macrophage, it is difficult to establish with certainty that these cells represent mature macrophages mobilized into the circulation. However, there is experimental evidence favouring the mobilization and circulation of mature macrophages.

Nicol and his colleagues (Helmy and Nicol, 1951; Nicol, Helmy and Abou-Zikry, 1952) showed in the guinea-pig that the injection of oestrogen results in a large increase in the number of vitally-stained macrophages in the liver and spleen, and the mobilization of hepatic and splenic macrophages into the circulation. In the liver and spleen of non-oestrogen-treated animals, a relatively small number of vitally-stained macrophages were seen. After oestrogen treatment the number of dye-bearing macrophages became considerably increased in both the liver and spleen and many of these appeared to be lying free in the splenic venous sinuses, splenic vein, and intralobular veins of the liver, but were not seen in the splenic artery. Nicol and Bilbey (1960) further studied the fate of hepatic and splenic macrophages which are liberated into the blood stream. They studied macrophage distribution following the intravenous injection of colloidal carbon in untreated mice and in mice treated with oestradiol monobenzoate. They observed that, two hours after carbon injection, the macrophages of the liver and spleen contained heavy concentrations of carbon, but, in marked contrast, no carbon was seen in the lungs, either in capillaries or in phagocytic cells. Twenty-four hours after carbon injection, the liver and spleen still contained large numbers of carbon-containing macrophages, while free macrophages containing carbon could also be seen in the splenic sinuses, splenic vein, portal vein, liver sinusoids, intralobular veins, hepatic veins, inferior vena cava, and right heart blood; but were not seen in the left heart blood, retro-orbital plexus, or femoral vein. Moreover, after 24 hours following carbon injection, increasing numbers of carbon-containing macrophages were seen in the alveolar walls, and also lying free in the alveoli, bronchi and trachea, and could be recovered in tracheal washings. In the animals treated with oestrogen the appearances suggested that the mobilization of splenic and hepatic macrophages and their elimination via the bronchial passages was an acceleration of a process which normally occurs after the intravenous injection of carbon.

Carbon injected intravenously is taken up almost exclusively by the intravascular macrophages, chiefly of the liver (about 90 per cent) and spleen (about 5 per cent), whereas when vital dyes are injected subcutaneously they become diffused throughout the tissues generally and are taken up by intravascular and extravascular macrophages. Nicol and Cordingley (1967) observed macrophage distribution for the 12 months following a single subcutaneous injection of trypan blue. They concluded that, behaving in the same manner as the intravascular macrophages, the free extravascular macrophages also migrate to the lungs, since, long after the liver and spleen became devoid of trypan-blue-containing cells, dye-bearing macrophages were present in considerable numbers in the lung washings. From these studies, it would appear that the bronchial tree may be an excretory pathway for intravascular and extravascular macrophages, and that the normal alveolar macrophage population may be partly derived or augmented from the liver, spleen and distant connective tissues. This view would not contradict the findings during experiments with chromosome markers that one-third of the alveolar macrophage population originates *in situ*, and two-thirds originate elsewhere in the body (Pinkett *et al.* 1966).

The above findings on the histological evidence of macrophage migration by Nicol and his colleagues should be considered in relation to the observations of Roser (1965, 1968) on the fate of injected labelled macrophages. Roser (1965) examined the fate of mouse peritoneal and alveolar macrophages which had been allowed to ingest radioactive colloidal gold particles in donor animals, and which were then injected intravenously into syngeneic recipient mice. It was found that with peritoneal or alveolar macrophages, radioactivity became localized initially in the lungs, but was then released slowly until after 8 hours less than 15 per cent remained, and the radioactivity accumulated largely in the liver and, to a lesser extent, the spleen. Autoradiography confirmed that the radioactivity in these organs was due to macrophages containing the radioactive colloid. The radioactive-labelled macrophages in the liver (presumed to be donor cells) of the recipient mice were morphologically identical to the unlabelled host Kupffer cells, and both were capable of ingesting intravenously injected iron (Russell and Roser, 1966). Kupffer cells labelled with intravenously injected colloidal [198]Au were separated from liver parenchymal cells by differential centrifugation and injected intravenously in syngeneic recipients. These cells were also temporarily retained in the lungs before labelled cells were detected in large numbers in the liver and, to a much smaller extent, in the spleen. These findings could be accounted for by the release of the radioactive label from the donor macrophages and its uptake by sinusoidal macrophages in the recipient mice, but studies on the retention of [198]Au by macrophages in cell-impermeable diffusion chambers *in vivo* indicated that, in the experiments quoted above, the label remained associated with the original injected cells (Roser, 1968). Thus the findings of Roser and his associates indicate that peritoneal and alveolar macrophages are capable of giving rise to Kupffer cells, and that peritoneal, alveolar and liver macrophages are also capable of giving rise to splenic macrophages.

Further long-term studies by Roser (1968) on the fate of injected macrophages were carried out using cells labelled with [195]Au which has a longer half-life than [198]Au. This enabled the fate of the injected cells to be examined over a much longer period, and showed that after the early localization of labelled donor alveolar or peritoneal macrophages in the lung and subsequent transfer to liver and spleen, the amount of label in the liver, spleen and lungs then remained constant for the eight-week period over which they were examined. In parabiotic mice in which one partner received labelled cells while the cross-circulation was occluded, very little radioactivity entered the unlabelled partner after the cross-circulation was re-established. These findings suggest that the fixed macrophages do not re-enter the circulation in any quantity over the duration of the experiments.

It would appear at first sight that the findings by Nicol and his colleagues, of sinusoidal macrophages of the liver and spleen migrating via the blood to the bronchial tree are superficially at variance with the findings of Roser that intravenously injected macrophages are largely sequestrated in the liver and spleen for a long period of time. Roser's studies suggest that the mobilization of sinusoidal macrophages may be a slow process when the cells are lightly loaded with colloid, but that this process may be accelerated when the cells contain larger amounts as in the experiments of Nicol and his associates. Moreover, the findings of the latter researchers indicate that the cells which were very heavily loaded with colloid were mobilized first, and that the complete replacement of colloid-containing macrophages in the liver and spleen by a fresh population of non-colloid-containing cells could take several months. Perhaps the apparent differences, possibly due in part to differences in the strains of animals used, could be resolved by injecting large quantities of carbon intravenously into mice which had previously been given intravenous injections of alveolar or peritoneal macrophages labelled with radioactive colloidal gold.

Vernon-Roberts (1969 *a*) injected washed carbon-containing peritoneal macrophages into the peritoneal cavity of recipient mice in which glass coverslips were inserted subcutaneously over the posterior thorax to initiate an acute inflammatory response. The recipient mice were killed at various intervals and the coverslips removed. Examination of the coverslips revealed that, although many carbon-containing macrophages were present at the inflammatory site 24 hours after injection, the greatest numbers were present at 48 hours. Microscopic examination of the liver did not reveal any carbon in the Kupffer cells during this period, indicating that no freed carbon had been released into the circulation to label the inflammatory macrophages in this way. These findings indicate that peritoneal macrophages may be mobilized to inflammatory sites situated elsewhere in the body. In further studies (Vernon-Roberts, unpublished results), it was also noted that macrophages containing carbon could be isolated in the peripheral blood; and, when carbon-labelled peritoneal macrophages were injected intraperitoneally or intravenously, there was a significant accumulation of carbon-containing macrophages in the lungs 72 hours after injection.

MIGRATION AND FATE OF INFLAMMATORY MACROPHAGES

There is a vast literature describing various aspects of the inflammatory process, but there is relatively little information available as to the ultimate fate of the mononuclear cells concerned in this phenomenon.

It is well recognized that the acute phase of inflammation is, in terms of the cells concerned, characterized by the early emigration of large numbers of neutrophil polymorphonuclear leucocytes together with a relatively small number of monocytes. If the inflammatory process progresses, the number of emigrating polymorphs is progressively reduced and the number of monocytes is progressively increased, so that, in most inflammatory sites of about 72 hours' duration, the mononuclear phagocyte is the predominant cell. However, depending on the nature and duration of the stimulus, polymorphs may continue to emigrate into the area for a very much longer period (at least several months) although they are usually present in much smaller numbers than the macrophages. In summary, it may be said that, depending on the inflammatory stimulus, acute inflammation is characterized by the predominance of polymorphs within the cellular exudate and in duration it may be transient (about 24 hours) or prolonged. Chronic inflammation always passes through an acute phase, and is characterized by the predominance of macrophages, but other cells such as lymphocytes, plasma cells and fibroblasts are also commonly present in variable proportions.

The studies of Spector, Walters and Willoughby (1967) on the emigration of polymorphs and mononuclears in chronic inflammation suggest that separate and distinct forces draw polymorphs and mononuclears into inflammatory sites. One possibility is that the attracting force for polymorphs wanes with time, whereas that which attracts mononuclears does not. Alternatively, the stimulus for emigration is the same for both types of cell, and the type of cell migrating into the area is dependent on the intensity of the stimulus. These events are complicated by the stimuli for the proliferation of mononuclear phagocytes within areas of chronic inflammation.

Attempts to demonstrate specific chemotactic factors for monocytes have generally proved unfruitful, whereas there are substances which appear to provide specific chemotaxis for polymorphs. However, recent studies suggest that serum may contain substances, probably fractions of complement (a substance necessary for the full activity of antibody on cells), which are specifically chemotactic for monocytes. It has been widely observed that the mononuclear phagocytes of inflammatory exudates avidly ingest the cytoplasmic fragments of polymorphonuclear leucocytes, and this has been suggested by Rebuck, Whitehouse and Noonan (1967) as indicating the ultimate use of such fragments in further mononuclear function, especially in 'mononuclear energization'. In this connection, Page and Good (1958) observed that the condition of agranulocytosis (the absence or extreme deficiency of neutrophil polymorphs in the blood) produced not only the absence of polymorph migration at the local inflammatory site, but failure of subsequent mononuclear migration as well. However, when these workers applied viable polymorphs to agranulocytic test lesions, normal mononuclear migration and function were restored at the local inflammatory site. These findings suggest that the presence of granulocytes at inflammatory sites may act chemotactically in some way to attract monocytes to the area.

Little is known about the fate of macrophages in acute inflammatory lesions which resolve rapidly. Examination of regional lymph nodes draining areas of acute and chronic inflammation suggest that many macrophages may migrate to the lymph nodes carrying with them dead tissue and foreign material which has been phagocytosed during the inflammatory process. Depending on the intensity of inflammation, some of the macrophages undoubtedly die and are taken up by other macrophages which migrate to the lymph nodes in due course. The presence of cells containing iron pigments resulting from haemoglobin breakdown suggests that some macrophages may persist at sites of trauma for many years, and in this way form a 'monument' marking the site where inflammation originally took place.

4

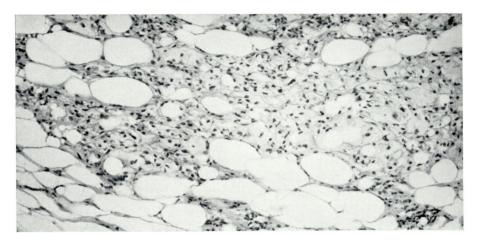

Fig. 18. Fat necrosis in human breast. Foamy lipid-containing macrophages ('lipo-phages') are present between the fat cells of the breast. Haematoxylin–eosin. × 140.

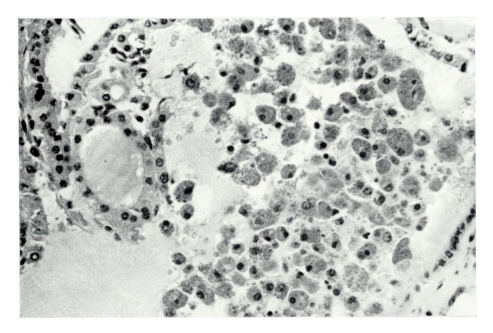

Fig. 19. Macrophages containing thyroid colloid in a diffuse colloid goitre. Haematoxylin–eosin. × 140.

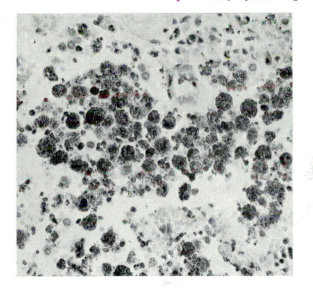

Fig. 20. Haemosiderin-containing macrophages ('heart-failure cells') in the lung in chronic pulmonary oedema. Haematoxylin–eosin. × 200.

The turnover and proliferative characteristics of macrophages in chronic inflammation have been discussed in a previous section. It would appear that some of the macrophages exhibit marked endocytic activity and are long-lived cells, whereas others exhibit much less phagocytic activity and have a short life-span. A variable proportion of these cells, depending upon the nature of the inflammatory stimulus, may transform to epithelioid and multinucleate giant cell forms which may themselves persist and proliferate for a very long time in the area. The transformation of macrophages into these cells, and possibly into fibroblasts, is discussed in the next section. The ultimate fate of the cells in chronic inflammatory lesions which resolve is largely unknown, but they probably die or migrate elsewhere as in the case of resolving acute inflammatory lesions.

In both acute and chronic inflammation, the cytoplasmic appearance of macrophages may change according to the pathological environment by phagocytosing materials present in the lesions. For example, in fat necrosis of the breast macrophages may have a very 'foamy' cytoplasm due to ingested fat (Fig. 18) which can be confirmed by the use of suitable stains. In various types of pathological processes, macrophages may exhibit unusual appearances by containing abundant pigment such as iron compounds, melanin, or foreign pigment material (tattoos); or they may contain abundant fat, cholesterol, sebaceous material, keratin,

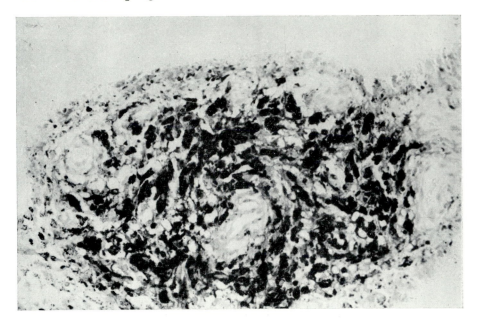

Fig. 21. Macrophages containing iron pigment in synovial tissues following haemorrhage into a joint cavity. Perls–neutral red. × 200.

thyroid colloid, cell debris or other material within their cytoplasm (Figs. 19, 20, 21). The identification of the ingested material can sometimes be a useful diagnostic aid to the practising pathologist.

TRANSFORMATION OF MACROPHAGES TO OTHER CELL TYPES

EPITHELIOID CELLS AND MULTINUCLEATE GIANT CELLS

Some of the evidence concerning the transformation of monocytes and macrophages to epithelioid cells and multinucleate giant cells *in vitro* has already been cited in the section dealing with the transformation of monocytes to macrophages.

Epithelioid macrophages and giant cells are characteristic features of infections characterized by the formation of granulomata. The granulomata are lesions which are classically, but not exclusively, seen in tuberculosis, syphilis and leprosy (the 'specific infective granulomata') and may have a central area of tissue necrosis with a surrounding zone of epithelioid cells and giant cells (Fig. 22). The epithelioid cells are enlarged macrophages having abundant cytoplasm, varying in shape but generally

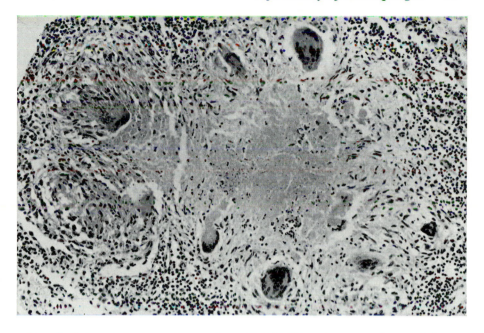

Fig. 22. Tuberculous granuloma in lung of child. There is a central area of tissue necrosis which is surrounded by a zone of epithelioid macrophages and multi-nucleate giant cells, and a peripheral accumulation of small lymphocytes. Haematoxylin–eosin. × 120.

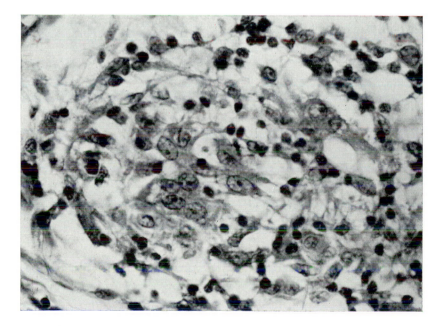

Fig. 23. Epithelioid cell granuloma in lymph node in sarcoidosis. Numerous cyto-plasmic processes extend from the cells. Haematoxylin–eosin. × 1,000.

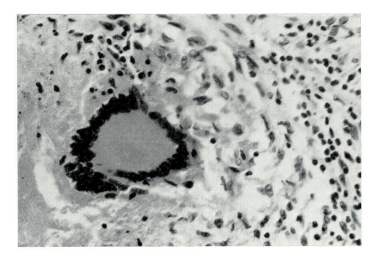

Fig. 24. Multinucleate giant cell in tuberculosis. Shows peripheral arrangement of numerous nuclei, and abundant cytoplasm. Haematoxylin–eosin. × 900.

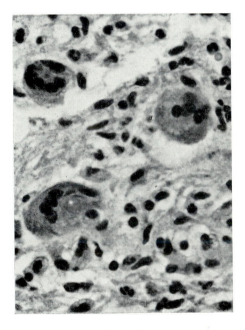 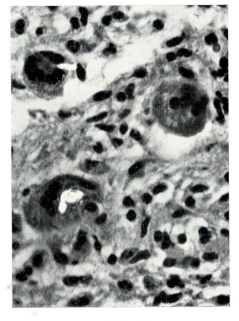

Fig. 25   Fig. 26

Fig. 25. Multinucleate giant cells and macrophages in tissue reaction to unabsorbed suture material. Haematoxylin–eosin. × 1,000.

Fig. 26. Same field as Fig. 25 photographed under polarized light. Shows birefringent (white) suture material within giant cells.

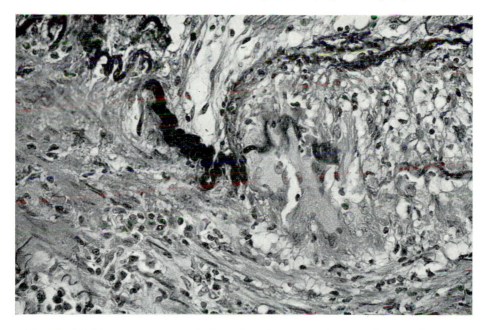

Fig. 27. Multinucleate giant cells ingesting fragments of internal elastic lamina (black) in the temporal artery in a case of giant cell arteritis. Haematoxylin–eosin. × 1,000.

polygonal, and have lightly staining rounded or oval nuclei. They have slim cytoplasmic extensions which are tightly entwined with adjacent epithelioid cells (Fig. 23). They contain many mitochondria, lysosomes and large vacuoles. The giant cells may contain between two and fifty or more nuclei which are generally situated at the periphery of the cell, either in circumferential fashion or grouped at one end of the cell (usually in the cytoplasm away from the centre of the granuloma) and contain abundant cytoplasm; the nuclear and cytoplasmic morphology is generally similar to that of the epithelioid cells (Fig. 24). Giant cells of identical appearance to those of the infective granulomata are also commonly seen in the reaction of the tissues to foreign material, such as unabsorbed sutures, implanted solid particles, injected oily solutions, and around ruptured cysts and damaged tissues and, under these circumstances, may represent the joining together of macrophages to deal with very large particles or to isolate material which cannot be readily absorbed (Figs. 25, 26, 27).

The presence of tubercle bacilli or foreign material in cultures is not a prerequisite to the formation of epithelioid cells and giant cells, as culture of pure suspensions of blood monocytes and peritoneal macrophages, and their subsequent transformation to epithelioid cells and giant cells, have clearly shown. However, it is true that giant cell formation is more common in cultures of whole blood, when the macrophages engulf the cell debris of erythrocytes and other leucocytes, than in cultures of peritoneal macrophages when much less debris is available. Moreover, the addition of tubercle bacilli to cultures of blood monocytes readily evokes the transformation of macrophages to epithelioid cells and giant cells which may progress to assume the topographical distribution of the classical tubercle follicle. Under the same circumstances, macrophages of lymph glands and spleen, and the Kupffer cells of the liver are also able to form epithelioid cells and giant cells.

The transformation of macrophages to epithelioid cells and giant cells *in vivo* is largely based upon circumstantial evidence. Giant cells have been observed in the spleen after the intravenous injection of large amounts of carbon (Patek and Bernick, 1960), and multinucleate giant cells are commonly observed in the tissue reactions to implants of foreign material, and at the site of injection and in draining lymph nodes after the administration of water-in-oil emulsions in common use as adjuvants.

Giant cells are regarded by some as being the products of the fusion of epithelioid cells, whereas the opposite opinion is that they arise by the amitotic division of epithelioid cells. Studies on the uptake of Kveim antigen in the presence of active sarcoidosis in humans (Hirsch, Fedorko and Dwyer, 1966) have shown that, when the Kveim antigen (extract of material from lymph nodes of a person known to have active sarcoidosis) is mixed with particulate colloidal gold or carbon, the particles are taken up by macrophages in the reaction. The particles are sequestered in phagosomes within the macrophages, and these cells are incorporated into the enlarging granuloma. Examination of epithelioid cells up to 60 days thereafter reveals the continuing presence of cells containing large amounts of the colloid marker, suggesting the persistence of these cells throughout this time. In these studies, the giant cells of the granuloma appeared to result from the fusion of pre-existing macrophages. Following fusion, the nuclei became oriented in a peripheral fashion and the cytoplasmic organelles were situated in the central region in a mixed arrangement. There was also an increase in mitochondria and other organelles.

The studies of Gillman and Wright (1966) also support the view that giant cells arise by fusion, and not by the division of macrophages. They gave rats injections of tritiated thymidine before or after implanting foreign bodies in the form of polyvinyl sponges or millipore filters, and studied the lesions using autoradiographs. The absence of label in the nuclei of the giant cells in rats which were labelled two hours before killing indicated that these cells so not synthesize DNA. In the rats labelled before the implantation of foreign material, 60 to 80 per cent of the mononuclears in the inflammatory site were labelled and it was thought that many of these cells were derived from blood monocytes, many of which were also labelled. In the multinucleate giant cells formed in this latter group, not all nuclei contained label, and it was concluded that these cells arose by fusion of emigrated blood mononuclears and some locally derived cells. Ryan and Spector (1970) have shown that DNA synthesis does occur in the nuclei of giant cells, providing that the cells are not less than 2–4 weeks old.

FIBROBLASTS

The evidence for the transformation of macrophages to fibroblasts is not conclusive. Both *in vivo* and *in vitro* studies have produced much evidence favouring such a transformation, but it is almost impossible completely to eliminate such factors as the contamination of macrophage cultures by some fibroblasts. Moreover, macrophages *in vivo* and *in vitro* often assume a spindle shape (Fig. 28), and are very similar in morphology (by light microscopy) to some fibroblasts, and it is thus important to establish that the cells considered to be fibroblasts are forming collagen as demonstrated by ultrastructural or biochemical methods. Few studies have satisfied the strict criteria for the transformation of macrophages to fibroblasts, but some interesting data is available.

Gillman and Wright (1966) gave rats tritiated thymidine before wounding, and found that labelled blood mononuclear cells, or their progeny, were still present in the fibrosing wounds 14 days after wounding. Labelled cells with the morphological features of fibroblasts were still present in the lesions long after the last isotope injection, and were considered to be derived from blood mononuclear cells, possibly either monocytes or lymphocytes. Shelton and Rice (1959) studied the formation of fibroblasts and collagen production, by electron microscopy and biochemical analysis, from mouse peritoneal cells in diffusion chambers in the peritoneal cavities of mice. Similar studies using diffusion chambers *in vivo* containing cells obtained from buffy coats and peritoneal suspensions in various species, including man, have been performed by others. These diffusion chamber studies must be interpreted with caution, since not only may the initial cell population be contaminated with some fibroblasts which may proliferate at the expense of macrophages, but the filters used in these chambers must have a pore size below $0.45\,\mu$ and be properly sealed to be completely cell impermeable.

There have been a large number of reports by workers using tissue culture techniques on the transformation of monocytes and macrophages to fibroblasts *in vitro*. The fact that some researchers have claimed that

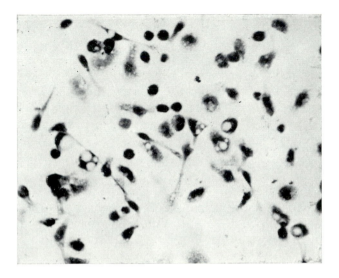

Fig. 28. Variable morphology of macrophages spreading on glass after 48 hours in culture. Many of the cells have assumed a spindle shape. Giemsa. × 700.

the process of transformation is the reverse, and that, in cultures of fibroblasts, macrophages appeared which seemed to have arisen from the fibroblasts, is indicative of the difficulties surrounding this type of investigation.

LYMPHOCYTES

The 're-utilization hypothesis' of lymphopoiesis was in part evoked to explain the long life-span of most lymphocytes. Trowell (1958) expressed the view that the lymphocyte is derived from a cell which is phagocytic in nature, and that macrophages ingest dead lymphocytes, assimilate (re-utilize) their nucleic acids, and proceed to differentiate into lymphocytes themselves. There is now a great deal of evidence which makes this theory of lymphopoiesis untenable. For example Mims (1962) showed that when peritoneal macrophages were incubated with thymidine-labelled lymphocytes, they ingested those lymphocytes which died, but did not incorporate the labelled nucleic acid into their own nuclei despite the fact that some of the macrophages underwent mitosis.

LYMPHOCYTE TO MACROPHAGE
TRANSFORMATION

Much data has accrued which indicates that cells having the morphology of the small lymphocyte can transform to macrophages. The demonstration of this development potential of some small lymphocytes has been achieved in experimental inflammatory lesions *in vivo*, and also by means of various *in vitro* techniques. However, the lymphocytic origin of some macrophages has not attained complete acceptance in all circles.

In the preceding section of this chapter dealing with the origin of macrophages, the possibility that a proportion of the precursor cells of monocytes and macrophages may have the morphological characteristics of small lymphocytes, and may circulate in this form, was examined.

The view that small lymphocytes transform to macrophages in inflamed areas has been expressed by many workers in the past. The most extensive evidence supporting this concept has been provided by the studies of Rebuck and his colleagues. Using the 'skin window' technique (Rebuck and Crowley, 1955), they have extensively examined inflammatory exudates in humans and described the apparent transformation of lymphocytes to macrophages occurring within 21 hours after lymphocytes had entered the area (about 3–6 hours after initiating inflammation). Within a short time of arrival at the inflammatory site, there was a gradual acquisition by the lymphocytes of nadi oxidase, peroxidase and alkaline phosphatase activity, and increasing numbers of sudanophilic cytoplasmic constitutents; and there was also increasing phagocytic activity (Rebuck, Monto, Monaghan and Riddle, 1958). In addition, individual living lymphocytes obtained from serial aspiration of stimulated cantharides blisters in man were continuously observed *in vitro* as they transformed into small monocytoid forms and macrophages, assuming a rosette of neutral red bodies, increasing in cytoplasmic volume, and also ingesting foreign material (Rebuck *et al.* 1958). Quantitative studies using the skin window technique showed that mitotic division of immigrant cells in the inflammatory exudates was not a significant factor in the accumulation of macrophages at the inflammatory site (Rebuck, Coffman, Bluhm and Barth, 1964). It was also shown that the lymphocytes in inflammatory exudates both store vital dyes and are metalophilic, and in this respect have similar properties to the macrophages (Rebuck, Boyd and Riddle, 1960). Rebuck has attributed the failure of many workers in the field of inflammation to elicit lymphocytic participation in inflammatory test lesions as being due to the absence of an applied antigenic stimulus at the site (Rebuck *et al.* 1964).

Opponents of the theory that lymphocytes transform to macrophages at inflammatory sites in skin windows claim that the precursor cells cannot be lymphocytes because lymphocytes are incapable of sticking to glass; however, quantitative studies on the stickiness of blood lymphocytes have shown that 11–70 per cent of uninjured blood lymphocytes from healthy donors may be retained in glass columns, and a higher percentage may be retained in the cells harvested from patients suffering from infections (Garvin, 1961; Johnson and Garvin, 1959).

Recent studies using radioactive isotopes to trace cells in inflammatory exudates have provided further evidence on lymphocyte to macrophage transformation. Volkman and Gowans (1965*a*, *b*) have suggested that some of the cells which originate in the bone marrow and are destined to appear as macrophages may, during their circulatory and migratory phases, resemble small lymphocytes rather than monocytes, but they conclusively demonstrated that the majority of small lymphocytes, which originate from lymph nodes, lymphoid germinal follicles and the thymus, cannot be antededents of macrophages. Further studies by Volkman (1966) showed that cells in peritoneal exudates which were morphologically indistinguishable from small lymphocytes were found to have the labelling features of a rapidly proliferating population, thus satisfying the criteria for the precursors of inflammatory macrophages, in contrast with the majority of small lymphocytes in blood and lymph which are slowly-produced long-lived cells. The experimental evidence indicated that the lymphocyte-like exudate cells had emigrated from the blood and that the bone marrow was the source of their precursors.

Vernon-Roberts (1969*a*) showed that, after a single subcutaneous injection of oestrogen in mice, there was an increase in the number of cells morphologically indistinguishable from small lymphocytes in the peritoneal cavity together with an increased percentage of these cells in the blood and peritoneal fluid which had incorporated tritiated thymidine during DNA synthesis. The increase in peritoneal lymphocytes was followed 24 hours later by an increase in the number of cells intermediate in appearance between lymphocyte and macrophage, and 48 hours later by an increase in the number of mature macrophages. These findings were taken to suggest lymphocytic origin for some of the peritoneal macrophages, since accompanying *in vitro* studies also showed that peritoneal lymphocytes undergo transformation to macrophages in culture, after passing through an intermediate cell stage.

Studies purporting to show lymphocyte to macrophage transformation *in vitro* have also been subject to criticism on the grounds that the initial cell suspensions were contaminated with monocytes or connective tissue cells. Bloom (1927) attempted to surmount this criticism by using cells from the thoracic duct of rabbits. The initial inoculum consisted predominantly of lymphocytes with small numbers of monocytes also present. In these cultures, lymphocytes apparently soon underwent transformation to macrophages, and in the process passed through a monocytoid stage. Hall and Furth (1938) also used thoracic duct cells but failed to demonstrate any transformation of lymphocytes to macrophages. These latter authors obtained their lymph by putting a needle into the thoracic duct, whereas Bloom had obtained lymph by incising the duct. Medawar (1940) also attempted to repeat Bloom's experiments, but observed no lymphocyte transformation in lymph obtained by thoracic duct cannulation, but she did observe the appearance of macrophages when she obtained lymph in the same manner as did Bloom.

The reason for the discrepancy in the results in the above experiments may be related to more recent findings in fluid culture systems. Gough, Elves and Israëls (1965) cultured an almost pure suspension of lymphocytes from normal human blood and found that after 12 to 24 hours macrophages made up between 20 to 30 per cent of the cells present. They concluded that most of the macrophages were derived from small lymphocytes since there were only about 5 per cent monocytes present initially, and no DNA synthesis, and therefore mitosis, could be detected using tritiated thymidine. In the same study, these workers labelled lymphocytes by incubation with ^{3}H-adenosine and demonstrated that macrophages which appeared in the cultures exhibited a grain count which corresponded closely with that of the initial labelled lymphocyte population. Gough and his co-workers also found that lymphocyte to macrophage transformation would not take place unless viable neutrophil polymorphs were added to the cultures. Vernon-Roberts (1969a) cultured almost pure suspensions of peritoneal lymphocytes and also found that a proportion of these cells transform to macrophages in the presence of viable polymorphs, but will not do so with killed polymorphs or in the absence of these cells. It is true that, in both of the experiments quoted above, the polymorphs, although the predominant cells added, may have been contaminated by other cells or factors which may have been responsible for promoting lymphocytic transformation. However, it is interesting to speculate about the role polymorphs or other added factors played in the original experiments of Bloom, Hall and Furth, and Medawar, since Medawar (1940) observed no polymorphs in the lymph obtained by thoracic duct cannulation, but there were 11.8 per cent polymorphs in the lymph which she obtained by Bloom's technique of incising the duct.

From these *in vivo* and *in vitro* studies, which only form a small proportion of the work done on this problem, it would seem that some small lymphocyte-type cells can transform to macrophages, possibly with the assistance of a factor provided by viable polymorphs. It is not possible at this time to express an opinion as to the extent to which this does or can occur *in vivo* until there is more definitive information available on the exact identity of the precursors of macrophages.

ONTOGENY OF MACROPHAGES

Macrophages appear very early during the development of vertebrates, and display increasing functional capacity in terms of the ingestion and disposal of various materials as embryonic life progresses. They appear long before cells which are capable of active antibody synthesis. Macrophages capable of ingesting vital dyes and actively engaged in erythrophagocytosis have been described in very early embryos in humans, chickens, rabbits, guinea-pigs, rats and cats (Maximow, 1932). They appear before erythropoiesis begins in the foetal liver, and appear to arise from undifferentiated mesenchyme and from primitive vascular endothelium. While continuing to arise from mesenchyme, they gradually assume the distribution found in the adult animal. Limited studies on the phagocytic activity of macrophages in embryos have shown that the macrophages of foetal vertebrates are less capable of digesting foreign material and destroying bacteria than those of adult animals, although phagocytic capacity and digestive abilities increase progressively with age (Karthigasu and Jenkin, 1963; Reade and Jenkin, 1965). Extensive clearance of injected materials does not take place in the foetus until there is functional activity within the liver and spleen, which are the organs primarily responsible for the clearance of circulating materials in the fully developed animal.

PHYLOGENY OF MACROPHAGES

The capacity for endocytosis is possessed to a greater or lesser degree by all cell types, and is generally related to cell nutrition. However, specialized cells with a highly-developed capacity for endocytosis and which play some part in the removal of micro-organisms and dead or damaged tissue appear early in evolution. Thus, Metchnikoff (1905) ascribed a defensive and nutritional role for certain phagocytic cells in starfish and sponges. In more complex invertebrates, such as the snail, octopus, earthworm, crayfish and bee-moth caterpillar, there are specialized phagocytic cells which will endocytose injected bacteria and foreign materials; in some, but not all, cases the phagocytes have the morphology of mammalian macrophages, and vary greatly in their organoid distribution in the individual species (Stuart, 1970). In all vertebrates, cells having the general morphology and functional capacity of macrophages and monocytes are readily identifiable in the blood and tissues.

Perhaps the most interesting problem concerning macrophage phylogeny is that there appears to exist a mechanism whereby specialized phagocytic cells can discriminate between foreign and effete material to be endocytosed and that which must not be ingested in those species which apparently do not possess the capacity to produce humoral antibody. However, there is evidence that in animals from coelenterates to molluscs, and probably the hagfish (a cyclostome), in which phagocytes also play some part in nutritional processes, humoral factors capable of reacting with foreign materials are present in unimmunized animals; moreover phagocytosis increases after immunization, although the formation of factors having the electrophoretic mobility of γ-globulins cannot be demonstrated (Nelson, 1969). Thus, although humoral factors facilitating phagocytosis exist in these animals, these factors are not comparable with the specific antibody which in higher species plays such an important part in the cellular recognition of foreign matter and in facilitating phagocytosis. Increased discriminatory capacity associated with the production of immunoglobulin antibodies and the possession of specialized monocytes and macrophages may be related to the development of a thymus, as in the lamprey (a cyclostome) and from elasmobranchs upwards (Nelson, 1969).

3

PHAGOCYTOSIS, PINOCYTOSIS AND VITAL STAINING

INTRODUCTION

Phagocytosis is the term used to describe the process by means of which cells are able to interiorize solid particulate matter. Fluid droplets are interiorized in much the same way by a process which is termed pinocytosis. In both cases, the ingested solid material or fluid is surrounded by a membrane formed by the interiorization of part of the plasma membrane of the cell; when the vesicle thus formed contains solid matter it is referred to as a phagosome, or if it contains fluid it is called a pinocytotic vesicle or pinosome. The mechanisms of phagocytosis and pinocytosis are probably identical, and differ only in the content of the vesicle formed. The term 'endocytosis' is sometimes used to denote both of these processes, and to emphasize that the basic mechanisms are fundamentally the same.

The presence of membrane folds seen in macrophages with the aid of the electron microscope has led to a hypothesis regarding a mechanism whereby substances adhering to the cell membrane are transported within the cell through sliding and vesiculation of this membrane. Inframicroscopic folds, which may be the result of the sliding of the membrane and invagination of the cytoplasm, became strangulated at their inner ends to form vacuoles. The walls of the vacuoles eventually break down and liberate substances contained in the vacuoles into the cytoplasm. This phenomenon can obviously only be seen with the electron microscope, but would help to explain the penetration into the cell of submicroscopic antigen substances or particles adhering to the cell membrane. The process has been given the name micropinocytosis or rhopheocytosis.

The electron microscope has also revealed that a variety of cells other than macrophages can also take up submicroscopic particles and fluid droplets in vesicles formed by invaginations of the cell membrane. These processes are usually concerned with cell nutrition, but pinocytosis may also be concerned with the transport of fluid across cells as in the case of capillary endothelium.

[64]

Polymorphonuclear leucocytes, fibroblasts, megakaryocytes, tumour cells, liver parenchymal cells, renal tubular cells, mesangial cells of renal glomeruli, pigment cells of the retina, Schwann cells, smooth and striated muscle cells, and possibly some epithelial cells, can display the ability to take up particulate matter under certain circumstances. The features which distinguish the macrophage from these other cells are not only its ability to take up large amounts of material and particles of relatively large size, but also its highly-developed ability to discriminate between particles of various types. The capacity of macrophages to distinguish between material to be phagocytosed and material not to be taken up is essential if they are to be effective functional units of metazoan organization. Apart from the fact that macrophages rarely ingest healthy auto-chthonous (host) tissues, there are many illustrations of the fine discrimination of which the cells are capable. Perhaps the most commonly occurring example of selective phagocytosis by macrophages is during normal haematoclasia when only effete or damaged erythrocytes are removed from the circulation. Moreover, possibly related to the fine discriminative ability of macrophages, there is evidence which suggests that the uptake of antigen by macrophages may be an essential step in the induction of some immune responses. It also seems likely that macrophages utilize their phagocytic ability in playing a part in the normal metabolism of iron and lipids.

In this chapter, the mechanism of phagocytosis will be considered in terms of the events taking place before, during and after the interiorization of material within macrophages. The mechanism of pinocytosis is considered separately. The kinetics of clearance of particulate matter from the blood stream, and the various factors which affect phagocytic rate and capacity will be considered in Chapter 4. The role of phagocytosis in immune mechanisms and in various metabolic processes is considered elsewhere.

FACTORS AFFECTING THE APPROACH OF MACROPHAGES TO MATERIAL FOR INGESTION

CHEMOTAXIS

Chemotaxis may be regarded as the first phase in the response of macrophages to foreign matter. It would appear that chemotaxis is irrelevant to the phagocytic activity of those macrophages which line the lumen of vascular and lymphatic channels as in the liver, spleen, bone marrow, lymph nodes and other sites where the particulate matter is conveyed to the phagocytes by the circulating blood or lymph. Chemotaxis may well be a factor in causing monocytes or macrophages to locate and emigrate into inflamed areas, and, having migrated into the tissues, the macrophages could conceivably be under the influence of the same or other chemotactic stimuli which induce them to move purposefully towards material to be ingested. Some experiments performed *in vitro* suggest that macrophages can respond to chemotactic stimuli. Harris (1953) recorded the movements of rabbit macrophages by the photographic tracing of their bright images during dark-ground microscopy. He found that the movements of the cells were random in control cultures. However, in cultures containing starch particles or various bacteria, the photographic images obtained indicated directional movement of the cells towards the particles. There was no directional movement to damaged tissue demonstrable in these experiments, although observations by others suggest that this may happen *in vivo*. Thus the apparent purposeful movement of monocytes towards erythrocytes individually damaged by a laser microbeam, and towards thyroid cells damaged or killed *in vitro* by cytotoxic antibody and complement, has been reported.

Initial attempts to demonstrate chemotactic responses using a special chamber to measure the emigration of cells through Millipore filters in response to chemotactic stimuli (Boyden, 1962) were successful with polymorphs but failed with macrophages. This was possibly due to the macrophages attempting to phagocytose the membrane itself. However, by using filters of larger pore size, Keller and Sorkin (1967) were able to demonstrate chemotaxis of rabbit peritoneal macrophages induced by casein and by a factor (specific for macrophages) present in some rabbit sera. Sorkin, Borel and Stecher (1970) have suggested that chemotaxis-inducing substances with a direct effect on cells should be termed 'cytotaxins', and those inducing the formation of cytotaxins should be denoted as 'cytotaxigens'. In an extensive investigation and review of agents chemotactic for macrophages and neutrophils *in vitro*, these authors reported that macrophage cytotaxins are found in normal sera, in plasma after incubation with antigen–antibody complexes, in guanosine and neutrophil granules, and in serum after incubation with neutrophil and liver post-granular fractions. A functional difference between the chemotactic response of alveolar and oil-induced peritoneal macrophages to the same stimuli was found, in that the peritoneal cells responded to casein, normal rabbit serum, and serum after interaction with the post-granular fraction from rabbit liver, whereas the alveolar cells did not respond at all to these factors.

When specific sensitizing antigen is added to suspensions of lymphocytes obtained from animals exhibiting delayed-type hypersensitivity to the antigen, a soluble factor which is not an immunoglobulin is produced which inhibits the migration of normal macrophages from capillary tubes *in vitro*. This factor is known as 'migration inhibition factor' (MIF) (David, Al-Askari, Lawrence and Thomas, 1964). Further studies have revealed that the sensitized lymphocytes also produce factors chemotactic for macrophages when cultured in the presence of specific antigen (Ward, Remold and David, 1969).

MECHANICAL FACTORS

Wood, Smith and Watson (1958) demonstrated that phagocytic cells operating in the parenchyma of the lungs are able to attack directly and ultimately destroy fully encapsulated pneumococci without the aid of an intermediary opsonin (a substance which facilitates phagocytosis). The ability of phagocytes to perform this function appeared to depend on the physical properties of alveolar and bronchial surfaces. Phagocytic cells operating on other tissue surfaces and upon relatively rough, inert material such as filter paper and fibre glass retained their ability to ingest virulent pneumococci, whereas on smooth surfaces such as those of glass, paraffin, albumin and cellophane, no phagocytosis took place, The authors suggested that the engulfment of bacteria which takes place in the interstices of inert fibre-like material *in vitro* in the absence of opsonins may be related to phagocytosis of bacteria in the network of fibrin which is frequently present in inflamed areas. 'Surface-dependent phagocytosis' as described above, was found only to operate on surfaces of suitable physical characteristics, whereas the presence of opsonins enabled phagocytes floating freely in a fluid medium to engulf the fully encapsulated organisms.

ATTACHMENT

It has now become apparent that the phagocytic process can be divided into two phases which appear to have separate determinants. The first phase is that of the attachment of the particle to the surface of the macrophage. The second phase is that of ingestion (engulfment) of the particle.

A clear separation of the attachment and ingestion phases has been achieved by various techniques which have demonstrated the attachment of particles to macrophages while the subsequent engulfment of the particles has been inhibited. Rabinovitch (1967*a*, *b*) showed that the attachment of glutaraldehyde-treated red cells to mouse peritoneal macrophages did not require the presence of serum factors or exogenous divalent cations and was less sensitive to ambient temperature than the ingestion phase. In contrast, the ingestion phase was initiated by serum or specific antibodies directed against red cell antigens or protein antigens bound to the red cells, and required the presence of divalent cations. It is possible to inhibit the ingestion phase by cooling cultures to temperatures below 20 °C, or by using metabolic inhibitors such as iodoacetate, even when antibody may be present.

CELLULAR RECOGNITION OF FOREIGN MATTER

Before the firm attachment of a particle to the cell membrane of the macrophage takes place, the cell has to discriminate between material which should be phagocytosed and that which should be left untouched. It appears that several mechanisms can play a part in enabling macrophages to 'recognize' materials which must be ingested for the benefit of the host organism. Available evidence suggests that the method of discrimination varies according to the nature of the particle, but in most cases probably involves antibody, either naturally-occurring or the result of immunization, cytophilic or (in the absence of antigen) non-cytophilic, and may also involve complement factors in some cases.

The term 'opsonin' was introduced at the beginning of this century to describe substances responsible for the heat-labile (56 °C) property of normal mammalian serum of rendering certain bacteria liable to phagocytosis by leucocytes. These substances were considered to be 'non-specific', since normal serum was found to promote the phagocytosis of a wide variety of antigenically unrelated bacteria. Subsequent work showed that the apparent heat-lability of the opsonins of normal serum is due to the fact that the presence of complement is necessary for their activity. It is now clear that there are no basic differences between the opsonins found in the serum of normal animals, and those which are present in immune sera. Both are specific, heat-labile (56 °C) substances, the phagocytosis-promoting effects of which are dependent on the presence of complement. The opsonins of non-immunized animals are, therefore, naturally-occurring antibodies.

Opsonins appear to be essential for phagocytosis to take place, and they do not merely enhance the phagocytic process which would take place at a somewhat slower rate in their absence. However, various types of particles are taken up by normal macrophages in the absence of any demonstrable humoral opsonic factors. At first sight this finding would suggest that macrophages might be capable of 'recognizing' certain kinds of immunologically unacceptable material without the involvement of specific antibody. Boyden (1966) has reviewed the evidence for the existence of naturally-occurring antibody, and has concluded that 'recognition' by the phagocytes of mammals is usually, if not always, mediated by antibodies in the serum or on cell surfaces, in the normal (non-immunized) as well as the immunized animal. The recognition by macrophages of immunologically 'inert' particles (such as carbon or gold) or damaged host cells would appear to require factors other than specific serum factors or cytophilic antibody. It may be concluded that the high degree of specificity in the discriminatory behaviour of phagocytes can only be accounted for by assuming that phagocytic discrimination is dependent upon differences in structure between 'familiar' macromolecules normally exposed to the body fluids and 'unfamiliar' macromolecules of foreign matter and of injured cells (Boyden, 1966). The evidence available at present also strongly suggests that the immune system similarly responds positively, by antibody production, or by the development of hypersensitivity, to any macromolecular configurations not normally exposed on the surfaces of healthy host cells or which are not normally present in the body fluids. In this connection, experimental immunization with damaged autochthonous (host) cells is generally successful in promoting the formation of antibodies (detectable by various means) against components of those cells.

Phagocytosis may follow the attachment of particles to macrophages by virtue of the presence of cytophilic antibody attached to the surfaces of these cells. Cytophilic antibodies have been defined as globulin components of antiserum which become attached to certain cells in such a way that these cells are subsequently capable of specifically absorbing antigen (Boyden, 1964). However, it appears that factors with the properties of cytophilic antibodies may be present in the sera of some animals not deliberately immunized with a particular antigen (Nelson, 1969). Moreover, there are circumstances in which cytophilic antibodies may be demonstrably attached to the surface of macrophages, but are demonstrable in the serum only with difficulty or not at all. The importance of cytophilic antibodies (either induced by immunization or naturally-occurring) in phagocytosis would appear to be primarily in the recognition of foreign matter by macrophages and may be the factors largely or wholly responsible for the fine discriminative ability of these cells.

Natural or immune antibodies which act as opsonins appear to function by promoting the attachment of particles to the surfaces of macrophages. The attachment could be promoted in the following ways: (1) the binding of particles to opsonin, forming an antigen–antibody complex, could occur prior to any contact with the cell, and the resulting complex could subsequently become attached to a receptor site on the surface of the cell; (2) the antibodies could be cytophilic and become attached to specific receptors for cytophilic antibodies (either $7S\gamma_2$, 19S or α_1 globulin cytophilic antibodies) enabling attachment of particle to antibody to occur at the cell surface; (3) antibodies which are non-cytophilic may be able to bind to specific receptor sites on the cell surface only after their configuration has been altered by previous contact with the antigen; and, (4) macrophages may possess receptors which may have specific profiles for binding with complementary profiles on material to be phagocytosed without the intervention of immune factors. There is evidence which suggests that all these factors may be operational in enabling macrophages to attach material to be phagocytosed.

Uhr (1965) suggested that antibodies which alone were not capable of attaching to macrophages became altered as a result of union with antigen so that a site was exposed which was similar to or identical with that on cytophilic antibodies. However, it is not clear whether complexes of non-cytophilic antibody and antigen attach to the same receptors as do antibodies which are cytophilic by themselves.

The role of complement in phagocytosis is controversial. Current opinion favours the view that complement is an accessory factor to the normal or immune opsonins which are essential for phagocytosis. It has been reported that antibodies to complement will block phagocytosis even in the presence of specific immune opsonins (Jeter, McKee and Mason, 1961). However, there is contradictory evidence regarding the necessity for the presence of complement during phagocytosis. For example, it has been shown that the ingestion of formaldehyde-treated erythrocytes by guinea-pig macrophages is largely unaffected, and possibly slightly increased, in the presence of added guinea-pig serum previously heated to 56 °C for 30 minutes to inactivate complement components $C'1$ and $C'2$ (Boyden, North and Faulkner, 1965). Moreover Mabry, Bass, Dodd, Wallace and Wright (1956) concluded that complement enhanced the rate, rather than the degree, of phagocytosis, and Bonnin and Schwartz (1954) found that if red cell antibodies were extremely potent, then erythrophagocytosis would occur in the absence of complement, but was enhanced when it was present. Patterson and Suszko (1966) found that the uptake of antigen–antibody complexes by macrophages from various species was not dependent on the presence of complement. Alternatively, complement could play a discriminatory role in the attachment phase of phagocytosis; thus complement could be fixed after the union of antibody with a particle and the resultant complex could become attached to a complement-specific receptor similar to the immune adherence receptor.

It has been reported that macrophages are capable of synthesizing at least one component of the complement complex βIC-globulin ($C'3$). Phillips and Thorbecke (1966) examined the origin of proteins synthesized in rat-into-mouse radiation chimeras using radio-immunoelectrophoresis, and found that peritoneal cell populations containing a high proportion of macrophages produced βIC-globulin, but suspensions of lymphoid cells containing few macrophages failed to produce this complement component. From these findings it is tempting to suggest that the variable reports regarding the requirement of complement in phagocytosis may be due to the ability of some macrophages to synthesize complement components, and that 'decomplementation' of added sera used in *in vitro* experiments does not necessarily deprive macrophages of some complement factors essential for phagocytosis.

'PIGGY-BACK' PHAGOCYTOSIS

The term 'piggy-back' phagocytosis has been introduced to describe a mechanism whereby certain substances, not normally ingested, may gain entrance to the cytoplasm of phagocytes (Sbarra, Shirley and Bardawil, 1962). This mechanism indicates that the discriminatory ability of these cells may be deceived under certain circumstances. These workers observed that malonate and fluorescent-labelled rabbit γ-globulin cannot enter a cell unless the cell is in the process of engulfing some other particulate material. They concluded that these materials, not normally ingested, enter the cell together with the particles within the membranous vesicle surrounding the selectively phagocytosed particles: the membrane was then thought to disintegrate, releasing particles and material into the cytoplasm.

ADHESION OF PARTICLES TO THE CELL SURFACE

In a preceding section, an outline has been given of the possible immunological factors concerned in the cellular recognition of foreign matter in terms of the possible receptor sites which may be present on the cell surface of macrophages. These postulated receptor sites for antigen-antibody complexes, cytophilic antibodies, non-cytophilic antibodies, complement, or non-immunological profiles may not only be concerned with the fine discriminatory ability of macrophages, but could also function as sites for the adherence of particles to the cell surface prior to ingestion. However, other less specific factors may also be operational in binding particles to the surfaces of macrophages.

The observations of Bangham and Pethica on zeta potential and calcium bridging may be relevant to the approach of macrophages to material for ingestion. If the approach of two particles or cells is considered, then the summative effect of adhesive energy forces (represented by London–van der Waals forces) and repulsive energy forces (represented by electrostatic forces) depends on the distance apart of the particles or cells. Moreover, there are a variety of other attractive forces which may be implicated in cell-to-cell adhesion (Pethica, 1961). It has been calculated that if a microvillus has a radius of curvature at its tip of about 0.1μ, two such tips could approach within 5Å with an interaction energy of 12 kT. The different adhesive properties between cells having the same overall energies of both attraction and repulsion may be refined down to a mechanism whereby local protruberances, having a small radius of curvature, enable close approach at ordinary thermal energies. Such a close approach might then result in successful bonding by calcium bridging, provided that the anion sites of the two types of cell were of the right type, e.g. carboxyl. The point contact would then spread to other areas of the cell surface (Bangham, 1964). Applying these observations to the approach of macrophages to material which must be phagocytosed, the macrophage is characterized in functional morphology by its ability to form abundant microvillous projections from its surface, and these microvilli can have a tip diameter below the critical $0.1\,\mu$ which, in theory, could allow a reduction in surface charge enabling the macrophage to approach the particle or cell close enough for calcium bridging to take place. There is supporting evidence in the necessity for a free supply of calcium ions for the attachment and engulfment of particles to take place.

It is well recognized that macrophages will readily adhere to a variety of surfaces. Special staining techniques indicate that macrophages have on their surface a layer of acid mucopolysaccharide, and this may be responsible for the adhesive properties of these cells. North (1966a) has also demonstrated that the outer components of the plasma membrane of guinea-pig peritoneal macrophages exhibit adenosine triphosphatase (ATPase) activity. The enzyme was active in the presence of calcium and magnesium ions, which are known to play an important role in phagocytosis. ATPase activity was inhibited by treatment with sulphydryl poisons or trypsin, and both these factors are also known to inhibit phagocytosis. Moreover, the adherence of phagocytic cells to glass is also inhibited by sulphydryl poisons. On the basis of his observations, North (1966a) concluded that, since plasma membrane movement is a necessary part of phagocytosis, and since the ATPases of other cells are known to act as mechanoenzymes (may function as an ATPase while undergoing structural changes), it would be reasonable to suggest that the cell membrane ATPase of macrophages plays a part in phagocytosis by utilizing phosphate-bound energy to perform membrane movement. The situation could broadly parallel the function of muscle myosin ATPase in muscular contraction.

INGESTION

The ingestion of particulate material by macrophages initiated a complex series of morphological and biochemical events within the cells. The morphological changes observed may vary slightly according to the situation of the cell and the type of particle being ingested, but the cytochemical events taking place within the cell after engulfment may be radically altered by the nature of the engulfed material.

MORPHOLOGICAL AND CYTOCHEMICAL CHANGES DURING INGESTION

Electron microscopy has revealed that an increase in electron density of the cell membrane may be seen at the site of attachment of a particle prior to ingestion. By using isolated cells it has been found that, if the particle to be phagocytosed is small, such as a single bacterium, the particle 'sinks' into the cell surface so that enclosure is largely effected by invagination. However, if the particle is relatively large, for instance a clump of bacteria, then enclosure is effected by the projection from the cell surface of microvilli which eventually surround the particle. In both cases, subsequent fusion of the distal ends of the invagination or microvilli results in re-formation of a continuous cell membrane, and also ensures that the ingested particle is contained in a vesicle bounded by a portion of the cell membrane which later becomes separated from the surface membrane by 'pinching off'. The vacuole formed by the pinching off of the cell membrane is termed a phagosome or phagocytic vacuole.

The formation of the phagosome initiates the digestive phase of intracellular events. As far as can be ascertained, the newly formed phagosome is devoid of hydrolytic enzymes (North, 1966*b*), and moves centripetally into the cell cytoplasm to interact with the lysosomes. It would appear that all the digestive enzymes in macrophages are segregated in cytoplasmic vesicles, the primary lysosomes, which are formed in the Golgi region and transport the contained enzymes to the vacuoles containing ingested material. In this way the digestive enzymes are always retained in a closed membrane system and do not, under normal circumstances, escape into the cytoplasm. There is abundant evidence that the phagosomes fuse with the pre-existing primary lysosomes in the macrophage cytoplasm, and the resulting phagolysosome or secondary lysosome contains both the ingested material and hydrolytic enzymes (Fig. 17). The fusion of lysosomes with phagosomes results in the depletion of pre-existing lysosomes associated with an increase in the appearance of enzymes (e.g. acid phosphatase) in the phagosomes. When the cells are actively taking up continuing numbers of particles, the absence of acid phosphatase in a large number of the phagocytic vacuoles together with its disappearance from the cytoplasm suggests that the supply of this enzyme is limited, possibly for a temporary period only (North, 1966*b*).

Apart from the changes taking place at the cell membrane during ingestion, and in the formation of phagolysosomes by the union of phagosomes and lysosomes with the subsequent depletion of the latter organelles and their enzymes, no other distinct morphological changes in cytoplasmic constituents have been observed by light or electron microscopy during the ingestion of particulate matter.

CHANGES IN METABOLISM

Macrophages exhibit marked changes in metabolism during and after the ingestion phase of phagocytosis. Dannenberg, Walter and Kapral (1963) postulated two phases in the metabolic changes accompanying the phagocytic process: they described an initial 'protoplasmic excitation' phase, brought about by the collision of the macrophage with particles and their ingestion, which results in an increase in oxygen consumption by the cells. Such phagocytes are 'activated' in that they subsequently ingest particles more readily and display increased motion and pseudopod formation. At this stage their enzymes are not increased in amount, but seem to be utilized to full capacity. The excitatory response by the cells initiates and merges with a phase of 'protoplasmic adaptation', which is a more lasting response on the part of the phagocyte to ingested material, and is expressed by an increase in certain enzymes and other changes in the cell. Thus, although the process of particle ingestion does not initially result in a quantitative increase in cell enzymes, the increased oxygen consumption and enzyme utilization indicates that the expenditure of additional energy is required during the event. Moreover, it can be demonstrated easily that, although attachment of a particle to the cell surface may have taken place, the ingestion phase of phagocytosis may be effectively inhibited by reducing metabolic activity, through lowering the environmental temperature or through the presence of metabolic inhibitors such as sulphydryl poisons (e.g. Mersalyl), iodoacetate, cyanide or dinitrophenol. The fact that metabolic inhibitors do not inhibit the attachment phase of phagocytosis indicates that it is unlikely that this process requires the significant expenditure of energy.

Evidence supporting the concept that alveolar macrophages are adapted to derive their energy by aerobic means mainly from oxidative phosphorylation, whereas peritoneal macrophages obtain their energy by anaerobic means from glycolysis, has been cited in Chapter 1. Consistent with this view is the finding that phagocytosis by alveolar macrophages is reduced under anaerobic conditions, whereas peritoneal macrophages can phagocytose normally during anaerobiosis. Oren *et al.* (1963) have demonstrated that guinea-pig peritoneal macrophages show far greater changes in metabolism during phagocytosis than do alveolar macrophages obtained from the same species. For example, the consumption of oxygen and the production of carbon dioxide from glucose were markedly increased in peritoneal macrophages when compared to alveolar macrophages, although there was a significant increase in these parameters with both cell types. Oren *et al.* (1963) also described the increased incorporation of ^{32}P into cell phospholipids during phagocytosis by guinea-pig peritoneal macrophages. This may indicate that some energy may be expended in the synthesis of cell membrane components to replace that portion of the cell membrane interiorized as the limiting membrane of the phagosome. The changes in lipid metabolism observed by Karnovsky *et al.* (1966) may also reflect synthesis of new membrane material during phagocytosis. They observed that, during phagocytosis, the incorporation of acetate and glucose carbon into lipids and the incorporation of ^{32}P into complex phosphatides was significantly increased. The results of both of the above groups of workers must be interpreted with caution since their findings were limited to peritoneal macrophages and have not yet been shown to occur in alveolar macrophages.

FATE OF INGESTED MATERIAL

Although previous workers had observed the presence of micro-organisms within cells, little attention was paid to this phenomenon until Metchnikoff brought to notice the particular ability of certain specialized cells to take up foreign matter and micro-organisms. About the same time, Pasteur established the importance of micro-organisms in the causation of disease processes, and Metchnikoff then made the fundamental contribution of realizing that the phagocytic cells which he had found to be present in many species were the major source of body defence against invading microbes. Much is now known about the susceptibility of various organisms to phagocytosis, and the fate of bacteria after their ingestion by phagocytes. A considerable literature also exists in connection with the fate of various types of micro-organisms and non-living materials after phagocytosis. It has become apparent that the fate of ingested material depends on the nature of the particle, and on factors such as the physical and chemical properties of non-living particulates, the type and virulence of bacteria, on immune factors such as opsonins, and on factors affecting the phagocytic ability of the macrophages, such as hormones, irradiation and certain cytotoxics.

BACTERIA

After ingestion by macrophages, bacteria can undergo one or more of the following processes: first, the bacteria may be acted upon by the enzymes with which the macrophage is abundantly endowed, and undergo death followed by digestion; secondly, bacteria may remain alive and, possibly, multiply within the cytoplasm of the cells; and, thirdly, the contact of the macrophage with the antigens of bacteria may set into motion a series of events culminating in the production of specific antibodies against those antigens by immunologically competent cells (not by macrophages).

Although a large number of studies have been performed *in vitro* which have shown that macrophages can inactivate a wide variety of gram-positive and gram-negative organisms, relatively little is known of the mechanisms which perform this function. Early workers put great emphasis on the acidic conditions which exist within macrophages after phagocytosis. It has previously been mentioned that phagocytosis causes an increase in metabolism within macrophages, and that this leads to the accumulation of lactic acid within the cell. It is postulated that, when the pH becomes sufficiently acid, lysosomal granules which have fused with phagosomes lyse and release their enzymes into the phagocytic vacuoles. Indicator dyes indicate a pH between 3 to 6 in the vicinity of engulfed particles. However, more recent evidence suggests that the acidity within phagocytic vacuoles does not play a major role in bacterial killing (Looke and Rowley, 1962).

The lysozyme (muramidase) of macrophages may play a part in the intracellular destruction of bacteria. Alveolar macrophages contain an abundance of this enzyme, but peritoneal macrophages are less abundantly endowed. Lysozyme is a low molecular weight protein which acts by hydrolysing complex acetyl-aminopolysaccharides, and is widely distributed in many tissues. The importance of lysozyme in intracellular killing is a matter of disagreement between various reports. However, there is little doubt that organisms which are susceptible to lysozyme *in vitro* are also killed after phagocytosis. For example, *M. lysodeikticus* is rapidly destroyed after phagocytosis by macrophages, but prior treatment of these organisms with methyl lysozyme protects them from lysozyme activity *in vitro* and from intracellular lysis (Glynn, Brumfitt and Salton, 1966). Moreover, strains of *M. lysodeikticus* which are susceptible to egg white lysozyme *in vitro* are rapidly killed after phagocytosis, whereas strains resistant to this *in vitro* treatment are highly resistant to intracellular killing (Brumfitt and Glynn, 1961).

The degree to which other lysosomal enzymes participate in the intracellular destruction of bacteria is largely unknown. In addition to lysozyme, macrophage lysosomes have been shown to contain acid phosphatase, lipase, cathepsin, acid ribonuclease, β-glucuronidase, aminopeptidase, succinic dehydrogenase, neuranimidase, hyaluronidase, aryl sulphatase and non-specific esterase. Macrophages also contain at least two proteases, one of which is a pepsin-like enzyme. Thus the macrophage is amply equipped to deal with a variety of substrates which may be presented to the cell as a result of the ingestion of bacteria. There is evidence which suggests that variations in resistance to bacterial infection may be related to changes in the activity of lysosomal enzymes. Thus it has been found that changes in phagocytic activity and resistance to infection may be paralleled by changes in the acid phosphatase activity of macrophages (Auzins and Rowley, 1962); but, although alveolar macrophages have a higher content of acid phosphatase, they display less bactericidal activity than peritoneal macrophages (Pavillard and Rowley, 1962). The latter discrepancy may be accounted for by the fact that a proportion of the lysosomal enzymes of the alveolar macrophages may not be discharged into phagocytic vacuoles but are already committed to utilization in other processes and are 'unavailable' for bacterial killing, as suggested by the findings of Leake and Myrvik (1966). Consistent with the view that the 'availability' of lysosomal enzymes is important in bacterial killing are the observations of Fauve and Delauney (1966). They demonstrated that mouse macrophages exposed *in vitro* to virulent bacteria capable of extensive intracellular multiplication (*M. tuberculosis*, *L. monocytogenes* and *B. melitensis*) enclosed the bacteria in small phagosomes, and there followed no reduction in the number of lysosomes possessed by the cells. In contrast, when the cells were exposed to non-virulent bacteria incapable of extensive intracellular multiplication (non-virulent *S. typhimurium* and *B. anthracis*) the bacteria were enclosed in large phagosomes and there was a marked reduction in the number of lysosomes present, indicating that lysosomal enzymes were released into the phagosomes during the process. These authors suggested that the differences in the lysosomal response of the cells to the virulent and avirulent bacteria might be determined by surface components of the ingested bacteria which result in differences in the osmotic pressure within the phagosomes. Whatever the mechanism involved, these findings clearly indicate an important role for lysosomes and their enzymes in the bactericidal activity of macrophages.

The digestion of bacteria by macrophages has also been studied with the aid of isotopically labelled organisms. Stähelin, Suter and Karnovsky (1956) examined the oxygen consumption and carbon dioxide production in ^{14}C-labelled virulent and avirulent bacteria. In the case of *M. tuberculosis*, the bacteria maintained their rates of oxygen consumption and labelled carbon dioxide production after phagocytosis. In contrast, avirulent *M. phlei* and *B. subtilis* both exhibited a decline in oxygen consumption and labelled carbon dioxide production after phagocytosis. Cohn (1963) studied the fate of *E. coli*, *B. subtilis*, *M. leisodykticus*, *Staphyloccus albus* and *S. typhimurium* labelled with ^{32}P or ^{14}C within granulocytes and macrophages. Both cell types brought about extensive degradation of bacterial lipids, nucleic acids and proteins. Intracellular breakdown was primarily dependent upon the composition of the ingested particle rather than on the type or source of the phagocyte. Following phagocytosis, labelled bacteria initially lost their pool of small molecular weight intermediates, suggesting an initial defect in the bacterial plasma membrane; and electron microscopy revealed that extensive fragmentation of the bacterial cell wall had occurred. This was followed by the degradation of acid-insoluble bacterial macromolecules with the formation of acid-soluble products. The majority of the bacterial breakdown products were then excreted by the cell into the surrounding medium. From an analysis of the degradation products, it appeared that bacterial protein, lipid and RNA were more readily digested than bacterial DNA. Heat-killed bacteria were more readily broken down than viable bacteria. The methods employed in this study did not enable the author to distinguish the fate of immunologically active components of the bacteria such as polysaccharides, but subsequent studies by Cohn (1964) showed that, although macrophages rapidly and extensively degraded the macromolecules of *E. coli*, there was no reduction in the antibody response to the *E. coli* agglutinogen obtained with *E. coli*–macrophage mixtures. In contrast, following the degradation of *E. coli* within polymorphonuclear leucocytes for the same period of time, the *E. coli* had lost 95 per cent of its antigenicity. These findings suggest fundamental differences in the outcome of the handling of bacterial antigens by the two types of phagocytes. The persistence of antigenicity following bacterial killing within macrophages is consistent with the possible role of these cells in the 'processing' of antigen as an initial step in antibody synthesis.

The importance of serum opsonins in promoting phagocytosis of bacteria by macrophages need no longer be stressed. There is additional evidence that serum opsonins also play a role in determining the fate of the intracellular bacteria. Rowley (1958) showed that mouse peritoneal macrophages which had taken up small numbers of *E. coli* in the absence of serum factors were unable to kill the ingested bacteria which multiplied intracellularly. However, if the same bacteria were treated with horse serum, rapid phagocytosis took place and the ingested bacteria were killed. Similar studies by Jenkin (1963) using virulent and avirulent strains of *S. typhimurium* showed that, in the absence of serum opsonins, bacteria which are ingested by mouse peritoneal macrophages survive and multiply intracellularly. It is of interest that, in Jenkin's studies, phage was used to investigate phagocytosis in the absence of opsonins. The finding that mouse serum that had been absorbed with bacteria was still capable of promoting phagocytosis providing the bacterium had first been coated with phage, suggested that phage plays an entirely passive role in the reaction between the bacterium and phagocytic cell. If, however, the serum was absorbed with phage-coated bacteria, then its phagocytic-promoting properties were lost. In the absence of any serum factors few bacteria were actually associated with the phagocytic cells, and the number of bacteria associated with the cells was not increased in the presence of phage. Thus the increased rates of phagocytosis observed in these experiments could not be due to phage particles by themselves. Jenkin visualized the conditions for optimal bacterial phagocytosis as being when the bacterium is not only coated with opsonins specific to the bacterium itself, but at the same time has a coating of phage which itself is coated with opsonins for the phage. The absence of phage and phage opsonins results in poor phagocytosis, even when opsonins for the virulent bacteria are present.

The role of complement in the intracellular killing of bacteria has been the subject of few studies. Most reports indicate that intracellular digestion is more efficient in the presence of fresh serum when compared to complement-free serum.

It thus appears that many factors may influence the fate of ingested bacteria within phagocytic cells. These include the anatomical situation of the cells, the type of bacteria, the presence of bacterial and phage opsonins, the metabolic state of the cell, lysosomal enzyme activity and availability, and possibly non-specific factors such as complement.

Finally, it must be borne in mind that certain bacteria are particularly capable of resisting intracellular digestion by macrophages, and this form of intracellular parasitism may play an important role in the pathology of diseases such as tuberculosis, leprosy, typhoid and brucellosis. The sequestration of bacteria within macrophages may also play a part in the spread of such organisms within the body, and may protect the organism from the effects of chemotherapeutic agents which may be present in the body fluids. The failure of macrophages to kill various types of bacteria may be related to immune deficiencies, deficiencies in macrophage structure or metabolism, specific properties of the particular organisms concerned, or to a combination of these factors.

VIRUSES

There is abundant evidence that macrophages play an important part in the pathology of some virus diseases by phagocytosing the virus particles. Some aspects of this process are the subject of an excellent review by Mims (1964). The clearance of viruses from the blood stream after intravenous injection have been studied to a limited extent. Among viruses whose clearance curves have been studied in mice are T_7 bacteriophage, poliomyelitis, Semliki Forest, vaccinia, Rift Valley fever (RVF), ectromelia, Newcastle Disease virus (NDV) and vesicular stomatitis. In general, the clearance curves obtained obey the criteria established for the clearance of particles from the blood by sinusoidal macrophages. The liver macrophages are quantitatively the most important cells responsible for the removal of viral particles from the circulating blood, and since the Kupffer cells are readily distinguishable from liver parenchymal cells, it naturally follows that the macrophage–parenchymal-cell aspects of virus infections of the liver have been particularly well studied.

When non-pathogenic viruses are injected intravenously, they are principally cleared by liver macrophages, and are subsequently digested and degraded. The fate of virus particles in liver macrophages has been studied using the fluorescent antibody technique. For example, the CL strain of vaccinia virus can be detected as antigen within Kupffer cells a few minutes after intravenous injection; however, within one hour after injection the fluorescence of the antigen vanishes and does not re-appear (Mims, 1964). It was concluded by Mims that most non-pathogenic viruses are destroyed by macrophages in this way. Viruses which are normally taken up by Kupffer cells and then go on to infect liver parenchymal cells may nevertheless be digested and destroyed by the Kupffer cells if they are premixed with specific immune serum. For example, when ectromelia virus is mixed with vaccinia immune rabbit serum and then injected intravenously into mice, the virus–antibody complexes, traceable with fluorescein-labelled antirabbit γ-globulin antibody, are taken up by liver macrophages, and there is no reappearance of ectromelia antigen in macrophages or hepatic cells (Mims, 1964). It was postulated that the normally infectious virus is perhaps rendered digestible by the specific antiserum. There is evidence that some viruses, RVF virus for example, may be taken up by liver macrophages without infecting them but then pass on to the liver parenchymal cells which subsequently exhibit rapid and extensive viral damage. It would appear that growth within macrophages, if it occurs, is not an essential part of this process.

It has been shown experimentally that in addition to the uptake of viruses by Kupffer cells after intravenous injection, viruses may also be taken up by macrophages in culture, by peritoneal macrophages *in vivo*, and by macrophages situated elsewhere in the body.

Despite the abundant evidence that macrophages are able to take up viruses by phagocytosis, it is clear that the penetration of virus particles into cells is not generally dependent upon this form of entry since receptors present on the cell surface specific to certain viruses probably play an important part in the ability of viruses to gain entry into cells. Once having entered a cell equipped for intracellular digestion, it appears that a virus can be destroyed by lysosomal enzymes released into the phagocytic vacuoles, although the virus may escape from the vacuole into the cytoplasm before digestion can take place. From studies performed with phagocytic cells other than macrophages, it would appear that lysosomal enzymes may be involved in the uncoating of the viral genome, which can then escape from the phagosome and initiate a replicative cycle within the cytoplasm (Silverstein and Dales, 1968). It would be interesting to know whether the same sequence of events takes place within macrophages after virus entry into a cell.

It has been established that macrophages are capable of synthesizing interferons. Interferons are cell proteins which act during the eclipse phase of intracellular viral multiplication and need the integrity of cellular RNA and protein synthesis. Interferons act through the derepression of anti-viral protein (actinomycin and puromycin block its action). The antiviral protein acts on viral protein synthesis at the polyribosomal level. Thus interferons protect cells by making other cells incapable of replicating viruses. Wagner and Smith (1968) have shown that large amounts of interferons are produced consistently at an extremely rapid rate in response to NDV infection of primary cultures of peritoneal macrophages. Adequate amounts are also produced 'spontaneously' (i.e. not in response to viral infection) by cultured macrophages, or after stimulation of these cells by bacterial endotoxins. The data obtained in these experiments were consistent with the synthesis of interferons by macrophages at comparable rates both with and without viral induction, and indicated that non-viral stimuli do not merely promote the release of preformed interferon or the activation of interferon precursors.

Recent studies by Allison (1970) suggest that the indefinite persistence of virus within macrophages of potential antibody-forming organs, such as spleen and lymph nodes, may be a relatively common phenomenon. This would provide an explanation for long-lived immunity against viruses, of greater duration than can be accounted for by memory cells of the lymphoid series. It is conceivable that small amounts of virus antigen could be repeatedly presented by macrophages to lymphoid cells so that small amounts of antibody could persist at a low level for an indefinite period of time, unless raised in response to re-infection or exposure to large amounts of virus released from another organ.

PROTOZOA, METAZOA AND FUNGI

Relatively little is known about the role of macrophages in the pathology of protozoal, metazoal and fungal diseases. Most of the available information concerns the protozoal infections caused by the genera *Plasmodium* and *Trypanosoma*. During malaria and trypanosomiasis, the causative parasites spend part of their life-cycle within macrophages. Moreover, infections with the genus *Leishmania* are characterized by parasitism and proliferation within macrophages. The majority of the experimental evidence favours a partly defensive role for the macrophages in these diseases, but this has not been firmly established.

Very little is known of the role of macrophages in response to metazoan infections. No all-embracing conclusions can be made since these organisms are composed of different tissues and organs which probably comprise many different antigens. A comprehensive review of antigen–antibody reactions in helminth infections has been carried out by Soulsby (1962).

6

The reaction of the body to fungal infections can vary considerably. In some instances, there may be very little cellular reaction by the tissues, but in most cases macrophages are involved in the response and can be seen to contain ingested fungi. Giant cell formation often ensues following the phagocytosis of fungi and there may be evidence of dead intracellular fungi within healed and active lesions. It thus appears that macrophages possess the capability of killing, and probably also of digesting, some fungi. However, the intracellular fate of fungi has not been investigated in any detail.

INORGANIC PARTICULATES

It is well known to histopathologists that the presence of macrophages containing pigments arising from the breakdown products of haemoglobin may be present in tissues many years after some form of trauma to the area visualized (Fig. 21). Likewise, it is widely known that increasing amounts of carbon and other dust particles are sequestered within macrophages of the lung especially in those individuals who live in cities or are occupationally exposed to the inhalation of dusts. Non-toxic particles (carbon, gold, diamond, iron compounds, chromium phosphate, etc.) remain enclosed within phagosomes and apparently do not affect the subsequent activity or fate of macrophages. In fact, under experimental conditions, the 'labelling' of macrophages with visible 'inert' colloids such as carbon is a useful means of tracing the subsequent activity and movements of these cells. However, silica is a notable exception to the generally harmless effects of phagocytosing inorganic materials by macrophages.

It has been well established that silica particles are intensely toxic for macrophages and kill them. After death, the macrophages containing silica disintegrate and the material is rephagocytosed. The cycle is repeated and a strong stimulus to fibrogenesis is exerted. Allison and his co-workers have shown that silica and asbestos particles liberate the enzymes from phagocyte lysosomes *in vitro*. However, aluminium-coated silica, or silica together with the protective agent polyvinyl pyridine *N*-oxide, were retained within the phagolysosomes of macrophages with little or no damaging effects on the cells; but phagolysosomes containing silica alone appeared to rupture, the lysosomal enzymes being released into the cytoplasm, after which the cells were damaged and some died (Allison, Harington and Birbeck, 1966). It has been postulated that the release of lysosomal enzymes is due to increased permeability of the phagolysosomes after hydrogen-bonding of silicic acid with active groups of the lysosomal membrane such as quaternary and phosphate ester groups of phospholipids, or, to a lesser extent, secondary amide (peptide) groups of proteins

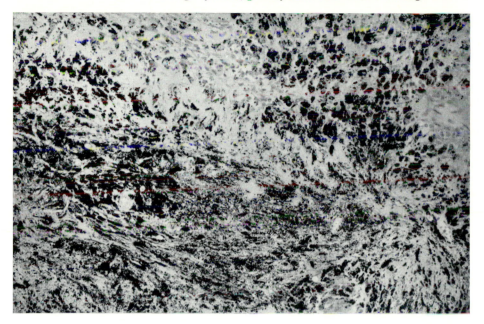

Fig. 29. Edge of silicotic nodule in lung of coal worker. Macrophages containing carbonaceous dust are present in the upper part of the micrograph. Scattered dust particles which have probably been liberated from disrupted macrophages are present in dense collagenous tissue in the lower part of the micrograph. Haematoxylin–eosin. × 120.

(Nash, Allison and Harington, 1967). Experimental evidence indicates that the phagocytosis of silica and the resulting disruption of macrophages is an essential preliminary for the accumulation of fibroblasts and the formation of collagen, and it has been suggested that the fibrogenic action of egg-white, agar, cartilage powder and carageenan may be initiated in a somewhat similar way (Curran, 1967). In connection with these experimental findings, it is of interest that there is abundant evidence that extensive fibrosis may occur in the lungs of coalworkers employed in mines where the silica content of the dust is high (Fig. 29). Moreover, the Caplan lesion, a vigorous chronic inflammatory process which takes place around silicotic nodules in the lung, may be seen in those who suffer from rheumatoid arthritis. The latter observation suggests that the genesis of silica fibrosis may also involve an immune mechanism in some way.

The experimental and clinical findings in silicosis have been favoured by some as indicating that silica may have antigenic properties. It has been found that 'silicotic hyaline' of silicotic nodules consists of 40 per cent collagen together with large amounts of α- and β-globulins, and it has been postulated that silicotic nodules are composed of collagen fibrils on which are deposited antigen–antibody precipitates (Vigliani and Pernis, 1958). These authors also believed that silica combined with serum protein to form an antigen, in response to which the body produced antibody over a long period of time.

In addition to silicosis, the phagocytosis of other inhaled particulates may also play an important role in the aetiology of other occupational lung disorders characterized by the development of pulmonary fibrosis. Macrophages are known actively to ingest inhaled particles of coal dust, asbestos, graphite, talc, iron oxide, haematite, kaolin, aluminium, barium sulphate, fibre glass and tin dioxide. There is evidence that a toxic effect on macrophages resulting in lysosome disruption, similar to the events seen after silica uptake, may initiate the fibrosis seen as a sequel to exposure to some of these dusts. Granuloma formation and fibrosis in the lungs may also be seen after the ingestion by alveolar phagocytes of protein-containing dusts of vegetable or animal origin, as in the case of Farmer's lung, baggasosis and byssinosis. A hypersensitivity type of reaction may be involved in the response to these and other organic materials.

EXOCYTOSIS

Exocytosis is the term used to describe the expulsion of material from within the cytoplasm of macrophages. It has been suggested, but not established, that macrophages may expel indigestible material, dense bodies and myelin figure formations. The ejection of erythrocytes and streptococci by monocytes has also been reported. The release of lysosomal enzymes from macrophages after phagocytosis *in vitro* has also been observed (Cohn and Wiener, 1963).

PINOCYTOSIS

Many investigators do not draw any distinction between phagocytosis and pinocytosis inasmuch as they are basically similar processes which differ primarily with respect to the quantities of liquid of the suspending medium that are taken up. However, pinocytosis may be a useful term to describe the means whereby inframicroscopic macromolecules and dissolved substances can enter cells when their entry cannot be accomplished by simple permeability processes. As in the case of phagocytosis, pinocytosis is not exclusive to macrophages and, with the aid of the electron microscope, evidence of pinocytosis may be seen in many cell types.

MORPHOLOGY AND MECHANISM OF PINOCYTOSIS

Holter (1959) has published an elegant review dealing with pinocytosis; and, although much of the review deals with pinocytosis in the amoeba, the observations are highly relevant to all mammalian cells. It has been shown that pinocytosis is a two-stage process. The initial step involves a reversible physiochemical adsorption of the dispersed substance to sites on the cell membrane. This phase is independent of temperature and of metabolic inhibitors. The second stage of uptake may be related to membrane synthesis and flow with the resultant incorporation of the adsorbed material into vesicles and vacuoles derived from the surface membrane. This stage is slower, irreversible, and sensitive to metabolic inhibitors and temperature.

There is suggestive evidence that the formation of a pinocytotic vacuole at the cell surface is induced by the prior fixation of protein to the cell membrane at that point, and that pinosomes cannot be formed in the absence of environmental protein in the medium surrounding the cell. This may be highly relevant to the role of pinocytosis in the ingestion of soluble antigens by macrophages (Robineaux and Pinet, 1960) where, presumably, discriminatory activity on the part of the macrophage would be expected to operate.

Morphologically, when viewed by light or electron microscopy, pinocytosis involves the fusion of microvilli around the fluid vesicle, similar to the process of phagocytosis, or the invagination into the cytoplasm of a small area of the cell membrane. In both cases, ingestion is completed by the fusion of the cell membrane around the fluid to form a membrane-bound vesicle or pinosome. Robineaux and Pinet (1960) have reported the migration of pinocytotic vesicles in macrophages from their point of origin at the cell membrane to the Golgi region where they subsequently progressively diminished in size by concentric reduction. The contents of the vesicles were ultimately absorbed into the cytoplasm. The same authors also reported that the pinocytosis of fluorescent-labelled protein by macrophages is eventually followed by the localization of fluorescence on mitochondrion-like structures in the Golgi region, and a more diffuse, but not homogeneous, fluorescence throughout the Golgi region. Ehrenreich and Cohn (1967) studied the uptake and fate of soluble proteins by macrophages using iodinated human serum albumin (HSA). It was found that HSA was taken up by pinocytosis and was eventually segregated within lysosomes. The rate of uptake was relatively slow when compared with the uptake of aggregated albumin. The HSA was rapidly degraded within the lysosomes, and subsequently the digestion product iodotyrosine was released into the surrounding medium. No other excretion products were detected. It was concluded that the digestion product iodotyrosine would not be expected to have importance in informative transfer in immune responses.

Cohn and his colleagues have shown that pinocytosis may play an important role in the formation of macrophage lysosomes (reviewed by Cohn, 1968). Cultures of macrophages exposed to low levels of serum in the medium formed few dense granules and little or no enzyme, whereas high levels of serum in the medium (50 per cent) resulted in the rapid and extensive production of lysosomes and lysosomal hydrolases. Examination of the cells exposed to high concentrations of serum by time-lapse cinematography revealed the origin of the lysosomes. Shortly after exposure to the medium the cells began to display active pinocytosis. Vesicles were seen to form at the cell membrane and streamed in uni-directional fashion into the perinuclear regions. Within a short time, large numbers of clear, phase-lucent, pinocytotic vacuoles had accumulated about the Golgi complex. Here, they underwent a transition in density, shrinking somewhat in the process, and became typical dense granules (lysosomes). Whereas the initial lucent pinocytotic vacuoles were uniformly negative for acid phosphatase, the dense granules were strongly positive, suggesting the acquisition of the enzyme during the formation of the dense granules. It was therefore apparent that the dense granules arose from pinocytotic vacuoles. The acquisition of acid phosphatase by the pinocytotic vacuoles was shown to be due to their fusion with Golgi vesicles (containing demonstrable acid phosphatase) in the Golgi region of the cell.

METABOLIC REQUIREMENTS

Cohn (1966) employed a microscopic assay to count the number of newly formed pinocytotic vesicles in the cytoplasm of macrophages after glutaraldehyde fixation. He used this technique to study the influence of agents which either induced or inhibited pinocytosis. The stimulus for pinocytosis in these experiments was 50 per cent newborn calf serum, which maintained pinocytotic vesicle formation at a high rate. It was found that pinocytosis was inhibited by (1) inhibitors of respiration including anaerobiosis, cyanide and antimycin A; (2) inhibition of oxidative phosphorylation by either 2,4-dinitrophenol (2,4-DNP) or oligomycin; (3) inhibitors of protein synthesis such as *p*-fluorophenylalanine and puromycin; and (4) inhibitors of glycolysis such as sodium fluoride or sodium iodoacetate. The fact that inhibitors of protein synthesis could block pinocytotic activity was taken to suggest that synthesis of new plasma membrane is probably required for continuing membrane interiorization. Additional studies with Actinomycin D indicated that pinocytosis was sensitive to this agent only after a lag period of about two hours. This suggested a dependance of pinocytosis upon DNA-directed RNA synthesis. The environmental temperature also played a critical role, and a 10 °C reduction to 27 °C resulted in a 70 per cent reduction in pinocytosis. Cohn concluded that pinocytosis is a temperature- and energy-dependent phenomenon in which protein synthesis is required. Moreover, the inhibition of pinocytosis by inhibiting oxidative phosphorylation suggests the ultimate participation of ATP in pinocytosis.

INDUCERS OF PINOCYTOSIS

Cohn and Parks (1967 *a*) studied various factors which stimulate macrophages to increased pinocytotic activity. In this case, the macrophages were cultured in the presence of 1 per cent newborn calf serum so that pinocytotic vesicle formation was minimal in the unstimulated cells. In general, vesicle flow was effectively stimulated by anionic molecules, whereas neutral and cationic molecules were much less effective and, in many instances, had no effect whatsoever. It was found that, of a number of proteins examined, the anionic bovine proteins plasma albumin and fetuin were potent stimulants, and were more effective than the cationic agents histone, protamine, and lysozyme. A similar stimulatory effect was obtained with a wide variety of anionic molecules, including homopolymers of glutamic acid, acidic polysaccharides and nucleic acids. Smaller negatively-charged agents, such as aspartic acid, glutamic acid and *N*-acetyl-neuraminic acid were also effective. The stimulant effect did not appear to be related to specific anionic groups on the molecules, but, in most cases, the minimum effective dose was inversely proportional to the molecular weight of the agent concerned. It was suggested that anionic molecules interact with the plasma membrane, perhaps with cationic phospholipid moieties, and somehow induce the energy-dependent process.

Further studies by Cohn and Parks (1967 *b*) demonstrated that adenosine, adenosine-5'-monophosphate (AMP), adenosine-5'-diphosphate (ADP), and adenosine-5'-triphosphate (ATP) were extremely potent inducers of pinocytosis. The 2' and 3' isomers of AMP as well as the nucleotides of inosine, guanosine, cytidine and uridine were much less effective. Of the nucleosides, only adenosine had any stimulatory effect. AMP, ADP, ATP and adenosine also caused increased spreading of the cells in culture accompanied by an increase in the number of spikelike cytoplasmic projections and an increased number of granules in the juxtanuclear region.

The most potent inducer of pinocytosis in macrophages has been found to be a bovine macroglobulin having the properties of an interspecies antibody (Cohn and Parks, 1967 *c*). The material is not present in foetal calf serum, but is present in progressively higher titres as cattle increase in age. In the presence of haemolytic complement, the macroglobulin haemolyses mouse erythrocytes and lyses mouse macrophages. The addition of the macroglobulin to foetal calf serum enhances pinocytotic activity, increases the number of lysosomes, and induces higher levels of acid hydrolases. Thus antibody directed against a macrophage membrane antigen stimulates pinocytosis, with the subsequent formation of increased numbers of hydrolase-containing lysosomes.

PINOCYTOSIS *IN VIVO*

From accumulated information, it would appear that pinocytosis by macrophages could be an essential mechanism for a variety of functions which may be performed by these cells.

Pinocytosis may be concerned in some way with cell nutrition, in that water and dissolved substances contained therein may be taken up by pinocytosis for this purpose. Pinocytosis has also been considered to play a part in the transport of fluid across cells, and it has been suggested that macrophages in certain sites, such as the liver, could be concerned in the transport of soluble proteins from the blood to the extravascular space (Hyman and Paldino, 1960), and thus play an important part in normal plasma protein turnover.

It has been established that pinocytosis is an important part of the mechanism of lysosome formation in macrophages. It is of interest to speculate how the findings of Cohn and Parks (1967 *a,b,c*) on various agents which stimulate pinocytosis may relate to those circumstances when an increased need for lysosomes may be required. There is abundant evidence that blood monocytes are the immediate precursors of the majority of macrophages which appear in early inflammatory lesions (see Chapter 2); and the monocyte is an immature macrophage in that it contains fewer lysosomes than the mature cell. From the studies of Cohn and Parks it is tempting to suggest that the presence of anionic proteins and AMP, ADP, and ATP in the fluid exudate may be the stimulus for the rapid maturation of monocytes to functionally mature macrophages by stimulating lysosome formation in the former cells. Moreover, the stimulation of pinocytosis in immature and mature macrophages could speed up the removal of noxious macromolecular and soluble materials by these cells.

Robineaux and Pinet (1960) have discussed the possible function of pinocytosis and micropinocytosis (rhopheocytosis) by macrophages in the uptake of antigen. There is little information available concerning the discriminatory ability of macrophages towards pinocytosable material. It has been reported that macrophages will take up soluble foreign proteins labelled by fluorescent or radioactive methods (Robineaux and Pinet, 1960; Ehrenreich and Cohn, 1967). This process will take place to a limited extent in the apparent absence of natural or immune antibody, but the addition of specific antibody results in a marked potentiation of the process (Sorkin and Boyden, 1959; Rhodes, 1964). Soluble antigen– antibody complexes are also taken up to a greater extent than antigen alone. Whether natural cytophilic antibody is involved in the uptake of foreign protein in the absence of serum factors is unknown. One may speculate that the immunological activation of pinocytosis observed by Cohn and Parks (1967*c*) could suggest that antibodies directed against membrane antigens of macrophages could facilitate the pinocytosis of antigenic molecules.

FATE OF PINOCYTOSED MATERIAL

The fate of pinocytosed material does not appear to differ very much from that of phagocytosed matter. Pinocytosed protein is segregated within macrophage lysosomes and is subsequently degraded (Ehrenreich and Cohn, 1967). Pinocytotic vacuoles may also disappear by concentric reduction in the Golgi region with the apparent incorporation of the con- tained fluid into the cytoplasm (Robineaux and Pinet, 1960).

REVERSE PINOCYTOSIS

Reverse pinocytosis is the means whereby macrophages (and other cells) can expel the fluid contents of vesicles from the cytoplasm. It is a normal feature of those cells in which pinocytosis occurs as part of a transport mechanism through their cytoplasm. In this case, the pinocytotic vesicles form chains of vesicles across the cytoplasm. The importance of reverse pinocytosis in macrophage function is unknown.

VITAL STAINING

Early investigations on the cellular components of the 'reticulo- endothelial system' were much enhanced by the introduction of vital staining techniques based upon Ehrlich's (1891) studies on the chemical constitution and cellular affinity of various dyes, and by the development by Ehrlich of relatively non-toxic colloidal dyes suitable for histological use. It was found by the early investigators that macrophages displayed a particular affinity for vital dyes and also were able to concentrate the dyes within their cytoplasm in a manner not exhibited by other cells which could take up the dyes to a limited extent (see review by Jaffé, 1938).

The dyes used for vital staining are negatively charged and most belong to the sulphonate triphenylmethane group of compounds, especially the diazo dyes. They have molecular weights between 600 and 1,000, but in water three to ten of these molecules aggregate together to form a micelle which has a much higher molecular weight. The diameter of the micelles is about 25 Å or below (Tanaka, 1961), which is much smaller than the inorganic colloidal particles commonly used to study the phagocytic activity of macrophages. In general (see Chapter 4), as the diameter of particles increases, the speed with which they are taken up by macrophages also increases. Thus most macromolecular particulates are taken up by macrophages within minutes or hours of exposure to the cells, but several days of exposure may be necessary for the cells to take up significant quantities of vital dyes.

Early studies using isolated perfused organs showed that vital stains are only taken up by macrophages after the stain has been bound to blood protein. More recently, Kojima and Imai (1961) carried out a detailed study of acid, neutral and basic vital dyes and have confirmed that they bind strongly to plasma albumin before being taken up by macrophages. Microscopy revealed that the dye–protein complexes were taken up by invagination and enclosure by microvilli. What is not certain is whether the vital dyes enter macrophages as a by-product of plasma albumin (to which the dyes automatically become bound after entering the blood stream) entering macrophages as part of an event which takes place under normal circumstances. In this connection, Kojima and Imai were of the opinion that vital staining may be considered to be a reflection (or tracer) of protein metabolism in which the macrophages are normally concerned. The evidence supporting a role of macrophages in normal protein transport and metabolism has been reviewed by Hyman and Paldino (1960).

Tanaka (1961) studied the mechanism of vital staining with the electron microscope and concluded that the uptake of vital dyes cannot be considered to be simple phagocytosis or pinocytosis. He suggested, on the basis of serial examination of cells, that dyes might enter cells by penetrating the limiting membranes of phagocytic or pinocytotic vacuoles or, more directly, by penetration of the cell membrane. In all cases, he was of the opinion that some of the dye granules became free in the cytoplasm. He described the material being segregated and condensed into non-specific pre-existing or newly-formed cytoplasmic structures which he termed 'segresomes'. The term segresome appears to be synonymous with the designation of the lysosome or autophagic vacuole. Some vital dyes (neutral red and Janus green) also became located within mitochondria. Thus Tanaka was of the opinion that vital staining can occur in association with phagocytosis and pinocytosis, but that these processes are not essential for vital staining to take place. Moreover, he concluded that the subsequent segregation of vital dyes involves a different mechanism to the digestive activity which usually follows phagocytosis and pinocytosis.

Repeated large doses of trypan blue or other vital dyes generally give rise to pronounced toxic effects. Nicol and Zikry (1952) observed that, in guinea-pigs given toxic doses of trypan blue, the toxic effects of the dye were much reduced by the administration of oestrogen, and the effect of oestrogen was much enhanced by castration in male animals. They attributed the effect to the hyperplasia and increased storage ability of macrophages throughout the body which they had previously observed following oestrogen administration (Nicol, Helmy and Abou-Zikry, 1952). Oestrogen has also been shown to have a protective effect against other toxic agents (Nicol, Vernon-Roberts and Quantock, 1965 a).

The prolonged administration of sub-toxic doses of trypan blue or Evans' blue can give rise to neoplastic proliferations of macrophages and the lympho-reticular tissues. Marshall (1956) showed that the subcutaneous administration of 1 ml of a 1 per cent solution of trypan blue or Evans' blue to rats at fortnightly intervals for 6 to 8 months could result in the production of neoplasms of the macrophages of the liver and, occasionally, of lymph nodes and lung. The initial changes occurred in the liver in the form of a monocytic infiltrate of the portal tracts together with proliferation and increased phagocytic activity of Kupffer cells. After 180 days of treatment, definite liver tumours appeared. Macroscopically such tumours were of two forms. In one type the tumour was a solitary well-defined mass in an otherwise normal liver. In the other, and more commonly observed type, the liver was grossly enlarged and studded with grey-blue nodules. The tumours were microscopically pleomorphic in appearance with multinucleate giant cells and cysts observed in some cases. The tumour cells exhibited a positive reaction to staining by the silver impregnation technique, and were thus established as being of the macrophage family. Evidence of metastatic spread was found in some instances. In addition to the well-defined tumour tissue, an intense diffuse proliferation of Kupffer cells accompanied by extensive erythrophagocytosis was observed. Marshall concluded that the hyperplastic and neoplastic proliferations of the macrophages produced by trypan blue and Evans' blue could not be the result of a prolonged non-specific 'irritation' of the cells by vital dyes, since other vital dyes (Vital red and Benzopurpurin 4B) or carbon particles do not produce similar changes. Moreover, the early proliferative lesions will retrogress if the dye treatment is stopped, and the advanced neoplastic lesions are apparently non-graftable into normal animals. Thus, the effect of the administration of trypan blue or Evans' blue is the induction of a graded neoplastic reaction in macrophages which does not differ fundamentally from the induction of neoplasia in other tissues with other carcinogens.

4

THE KINETICS OF PHAGOCYTOSIS. STIMULATION AND DEPRESSION OF THE PHAGOCYTIC ACTIVITY OF MACROPHAGES

INTRODUCTION

The cells which make up the macrophage system are scattered throughout all the tissues of the organism and *in vivo* techniques for the quantitative assessment of the phagocytic activity of all these cells at one and the same time are not available at present. However, the fact that a large number of the body's macrophages are situated in contact with the circulating blood enables the phagocytic activity of this group of macrophages to be assessed in quantitative terms by determining the rate at which these cells remove intravenously injected particulate matter. Although early investigators had made some attempts to assess the rate of clearance of intravenously injected materials, it was not until the kinetics of clearance of particulates were exhaustively studied by Halpern and his co-workers (reviewed by Benacerraf, Biozzi, Halpern and Stiffel, 1957) and by Dobson and his colleagues (reviewed by Dobson, 1957) that accurate data were first obtained. From these studies on clearance dynamics a number of clearance tests for the quantitative assessment of phagocytic activity have been evolved and are widely used today under experimental and clinical conditions. A limited number of studies have also been carried out of the quantitative assessment of macrophage phagocytic activity *in vitro*.

THE PHAGOCYTOSIS OF INTRAVENOUSLY INJECTED PARTICLES

CHOICE OF PARTICLE

The following criteria govern the choice of substances suitable for use in studying the phagocytic activity of macrophages in contact with the blood:

    1. The particles should only be taken up by macrophages in contact with the blood.

2. The particles should not cross the capillary barrier during the test period.
3. The particles should be homogeneous in size.
4. The preparation must be stable in the blood in the concentration used.
5. The substance should be non-toxic for macrophages and the host organism.
6. The substance should be stable *in vitro* and *in vivo* and should not be changed physically or chemically by contact with the blood.
7. The substance should be accurately measureable in the blood and in the tissues by chemical or physical methods. It should be traceable histologically within macrophages.

A wide range of test substances have been found to satisfy all or most of these criteria. They include particular suspensions of carbon, saccharated iron oxide, chromium phosphate labelled with ^{32}P, heat-aggregated serum proteins labelled with ^{131}I or ^{125}I, colloidal ^{198}Au, nucleated pigeon red blood cells, lipid emulsions, killed bacteria labelled with ^{51}Cr, ^{131}I or ^{32}P, thorotrast, silver iodide, methyl methacrylate and polystyrene latex labelled with ^{125}I.

The most commonly used of the above test particles are colloidal carbon in the experimental animal, and heat-aggregated human serum albumin labelled with ^{131}I or ^{125}I or lipid emulsions in man. Carbon is obtainable in a well-standardized non-toxic colloidal form, its concentration in the blood is easily assessed by measuring the optical density of haemolysed blood samples, and it is easily visualized in the tissues without the need for special staining techniques or autoradiography. Heat-aggregated human serum albumin labelled with radioactive iodine has been successfully used to study phagocytic activity in man (Ilio and Wagner, 1963). Large doses of this material have been administered repeatedly over long periods without evidence of hypersensitivity or excessive radiation. Phagocytic activity in animals and man has also been extensively investigated using intravenously injected lipid emulsions by DiLuzio and his co-workers. In more recent studies, the clearance of an emulsion composed of safflower oil, glycerol and lecithin was assessed in twenty-four normal human subjects by measuring serum triglyceride levels in the blood samples (DiLuzio and Blickens, 1966). No ill-effects were observed. Radioactive colloidal gold has also been used frequently for the assessment of phagocytic activity in man. Although no ill-effects have been reported, there are ethical objections to its use since it is well documented that the material is retained indefinitely within macrophages of the spleen and bone marrow, and the long-term effects of sequestering such colloids in these tissues are unknown.

DYNAMICS OF CLEARANCE OF PARTICLES FROM THE BLOOD

Particulates which are removed from the blood exclusively by macrophages following intravenous injection display a characteristic type of disappearance curve. With most particulates, three phases may be distinguished in the disappearance curve. The initial phase is due to the mixing of the injected particles with the blood. During the second phase the particles are removed from the blood in exponential fashion by the macrophages of the liver and, to a lesser extent, the spleen and other organs. The third phase indicates the presence of a more slowly disappearing fraction of smaller particles forming the 'tail' of the curve. The first and third phases may vary according to the size and dose of particles, but the second exponential phase has certain constant relationships which enable the phagocytic activity of the macrophages to be measured with a considerable degree of reproducible accuracy. The exponential shape of this portion of the curve reflects the progressive increase in the satiation of the macrophages by the ingested particles. If the ingested particles did not satiate the cells, the curve of clearance of particles from the blood would obviously continue as a straight line.

Of the test substances which are suitable for the quantitative assessment of clearance rates, carbon is the most widely used in animal experimental work. It is of course unsuitable for use in humans since the pigment is retained in the tissues for a very long time. Early studies on the clearance of colloidal carbon (Indian ink) from the blood were vitiated by the toxic reactions caused by the shellac present in the suspensions, in spite of the fact that stable homogeneous and well-dispersed suspensions of carbon were used. The shellac caused the liberation of thromboplastin resulting in the *in vivo* formation of microthrombi and flocculates of carbon which were arrested in the lungs (Biozzi, Benacerraf and Halpern, 1953). However, colloidal carbon suspensions free from shellac (Gunther Wagner, Pelikan Werke, Hanover, carbon code number C11/1431a) preserved in fish glue and stabilized with gelatin were found to be non-toxic, and affected neither the phagocytic activity of macrophages nor blood-clotting mechanisms (Biozzi *et al.* 1953). Examining the clearance of carbon after intravenous injection, Biozzi *et al.* (1953) made a study of the mathematical relationship between the speed of phagocytic activity, the dose of colloid injected, and the weight of the organs (liver and spleen) principally responsible for the phagocytosis of particles. The mathematical relationships derived from this study of the clearance of carbon are still in common use.

When colloidal carbon is injected intravenously in rats in doses between 4 mg and 48 mg per 100 g body weight, the clearance of carbon from the

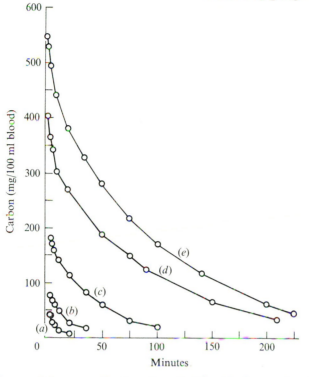

Fig. 30. Curves of clearance of carbon from the blood in the rat after the intravenous injection of single doses given on the basis of (a) 4 mg, (b) 8 mg, (c) 16 mg, (d) 32 mg, (e) 48 mg per 100 g body weight. Reproduced with permission from Biozzi *et al.* (1953).

blood is regular and progressive (Fig. 30). If the logarithm of the concentration of carbon is plotted against time (Fig. 31), the curves of clearance of carbon with respect to time are straight lines except for the highest doses of carbon, 32 mg and 48 mg per 100 g body weight. In these latter doses there is a straight line relationship after the first few minutes of clearance. From these findings, it can be generally stated that the concentration of a carbon suspension in the blood after injection is an exponential function of time, and can be expressed by the equation $(\log C_o - \log C)/T = K$, or $C = C_o^{-KT}$, where C is the concentration of carbon at time T, and C_o the blood concentration of carbon just after injection and before the particles are ingested by the macrophages. K is a constant which characterizes the rate of clearance of carbon from the blood and is called the 'total body phagocytic index' or the 'K value'.

If the mean K values obtained for each dose of carbon are plotted against the corresponding doses of carbon injected, a curve is obtained which is very close to a rectangular hyperbola. This means that the K value

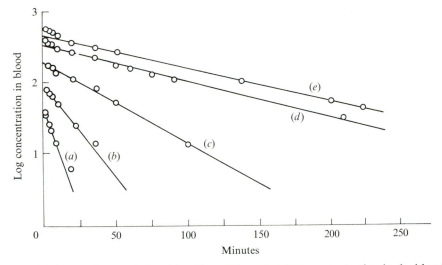

Fig. 31. Shows linear relationship of logarithm of carbon concentration in the blood when plotted against time, using data from Fig. 30. Reproduced with permission from Biozzi *et al.* (1953).

is inversely proportional to the dose of carbon injected. Thus $K \times D =$ Constant, where D is the dose of carbon injected. The observation that $K \times D$ is a constant indicates that, regardless of dose, the initial rate of carbon removal is constant. Such a relationship is not consistent with exponential clearance, since the exponential function is a description of a process in which the rate of removal is proportional to the amount present. Therefore the rate constant K should theoretically be independent of concentration instead of inversely proportional to it. This contradiction between exponential clearance and a rate constant which is dependent upon the injected dose has been explained by Benacerraf *et al.* (1957) as being due to the progressive saturating effect of the ingested particles decreasing the efficiency of clearance being counteracted by the increasing efficiency with which the Kupffer cells extract particles from the circulation as the blood concentration decreases. These findings invoke the concept of a constant phagocytic rate which yields linear instead of exponential kinetics, and the concept of gradually developing progressive 'blockade' of the macrophages.

Recent evidence obtained in studies on blockade (Dobson, Kelly and Finney, 1967) suggests that the kinetics may be more complex than the concept evolved from the observations of Benacerraf *et al.* (1957). Applications of mathematical models to the study of phagocytosis of intravenously injected particles (Fred and Shore, 1967) suggest that it may be invalid to accept previous conclusions that colloid clearance follows first order kinetics, with rate constants that are dependent upon initial concentration. Short segments of model clearance curves may appear to be first order (linearity of the semilogarithmic plot) when in fact they may be zero order. Moreover, analysis of model behaviour suggests that at low and intermediate concentrations of carbon, the macrophage system does not exhibit its maximum functional capacity which is measured by high doses alone. The experimental and mathematical findings of Fred and Shore (1967) strongly support the principle that an interaction at the macrophage cell membrane constitutes the rate-limiting factor in the phagocytosis of circulating particles. The mathematical model proposed by Gabrielli and Snell (1965) provides additional information agreeing with this concept. They interpret the early rapid phase of clearance as being due to adsorption of the test particles on to specific available sites on the cell surface. The process ends when all the sites are occupied. The occupied sites are ingested with the particles, and thereafter the rate of clearance is dependent upon the rate of regeneration of adsorbing sites.

When the dose, D, of carbon is reduced below a certain level, all the carbon is removed from the circulation by the macrophages of the liver. Thus below a critical dose level the observed K value is constant since the rate of clearance of the particles is purely a reflection of liver blood flow. This value (expressing liver blood flow) is referred to as K_{max}, and the critical dose is established according to the equation $(K \times D)/K_{max} =$ critical dose. When the dose of injected colloid exceeds the critical dose, the particles are not all removed from the blood in their first passage through the liver, and some are removed by splenic macrophages and, to a lesser extent, the macrophages lining sinusoids in other organs.

In practice, during investigation of phagocytic activity, blood samples are withdrawn at short intervals following the injection of carbon, and the concentration of carbon is measured by means of an absorptiometer. The K value (phagocytic index) is most conveniently calculated by the method of least squares as the regression coefficient of the straight line relating the logarithms of the absorptiometer readings plotted against time. The K value is not, however, a truly linear index (i.e. it is not directly proportional to the half-time of clearance) and it is more correct to express phagocytic activity as the reciprocal (K_r) of K, since this gives a linear arithmetic relationship $(K_r = 1/K)$ and allows a statistical analysis of the results to be made.

It is quite clear that the macrophages of the liver play a major role in clearing particles from the blood. It has become apparent that the spleen may also play a significant role depending on the dose of colloid.

ROLE OF THE LIVER AND SPLEEN

The Kupffer cells of the liver take up about 90 per cent of injected colloid during normal clearance tests. The role of the spleen is less important but, in general, the percentage recovered from the spleen increases with the dose. Thus, for very small doses of colloids, which are rapidly removed from the circulation, the Kupffer cells take up nearly all the injected colloid; however, with higher doses which may remain circulating for hours, the spleen contains more than twice as much colloid per gram of tissue as the liver (Benacerraf *et al.* 1957).

Different animal species show considerable variation in the rate of clearance of a fixed dose of colloid per unit of body weight and the approximate half-times of clearance of chromium phosphate are as follows: mouse, 30 seconds, rat, 45 seconds, rabbit, 40 seconds, chicken, 35 seconds, dog, 1.5 minutes and man, 2.5 minutes (Dobson, 1957). These differences in removal rates are largely due to differences in liver-to-body-weight ratio. Relative liver size alone however does not completely account for the differences in phagocytic rate. The relative size of the blood volume is also a factor because the disappearance rate constant (K) is directly related to the volume of blood perfusing the liver per minute. Moreover, the livers of some species have a higher perfusion rate, in that they have a higher blood flow per gram of tissue per minute. There is evidence which suggests that a higher perfusion rate may compensate for a smaller liver, thus providing an advantage in weight economy in some species (Dobson, 1957).

Benacerraf *et al.* (1957) formulated an equation for the relationship between the phagocytic index (K), the body weight (W), and the combined weights of the liver and spleen (W_{ls}), expressed as α (the corrected phagocytic index) $= W/W_{ls}.\sqrt[3]{K}$.

The corrected phagocytic index α, may be used as an expression of phagocytic activity when experimentally induced stimulation or depression of phagocytic activity is accompanied by changes in the weight of the liver and spleen. Its validity is exemplified by the observations of Benacerraf *et al.* (1957) that for a dose of carbon of 8 mg per 100 g bodyweight, the average value of K in the mouse was 0.047 and in the rabbit 0.008. The mouse has a relatively larger liver than the rabbit, and for the same doses of carbon, the average α value in both the mouse and the rabbit was 5.4. Although the corrected phagocytic index α of Benacerraf *et al.* (1957) is more commonly used, Di Carlo, Haynes and Phillips (1963) proposed a different equation for relating K values to changing weights of liver and spleen. Their corrected phagocytic index K_{ls} is derived from the formula $K_{ls} = 10^3\, K.LS_c/LS_e$, where LS_c and LS_e are the combined liver and spleen weights in control and experimental animals respectively.

EFFECT OF PARTICLE SIZE, NUMBER AND SURFACE CHARACTERISTICS

Following the intravenous injection of particulate matter, the clearance curve obtained generally exhibits a slowly disappearing component or 'tail'. It is now well established that this 'tail' portion of the colloid disappearance curve is due to the smaller particles which are relatively inefficiently phagocytosed by macrophages. Within limits of maximum size, the larger the particles the more rapidly will they be ingested by the macrophages. This can be amply demonstrated by injecting a suspension composed of particles of various sizes, and observing that the smaller particles remain circulating for longer periods. These findings are not due to differences in the number of particles injected, since it has been established that a saturating effect of the quantity of material already ingested by the cells is not a significant determining factor; but it seems likely that the rate of phagocytosis is predominently dependent upon the number of free particles yet to be phagocytosed, in addition to the size of the individual particles.

The size of the particles also greatly affects their subsequent distribution. Larger particles, which disappear relatively rapidly from the blood, tend to localize specifically in the macrophages of the liver and spleen. The slowly disappearing smaller particles localize to a high degree in the bone marrow (30 to 50 per cent). This phenomenon is not related to the total number of particles injected.

These observations stress the importance of using suspensions of particulates which are homogeneous in terms of particle size and number in quantitative assessments of phagocytic activity.

The surface characteristics of colloids may affect their clearance from the blood and their distribution in the various organs. Wilkins and Myers (1966) have demonstrated that colloids having different electrophoretic mobilities exhibit different patterns of organ distribution. Negatively charged colloids are largely taken up and retained in the liver, while positively charged colloids show an initial appreciable accumulation within the lungs and a later accumulation in the spleen. The fact that all such particles interact with serum and appear to adsorb a coating of some component which gives them the same electrophoretic mobility, suggests that the surface properties of a colloid may alter the steric arrangement of adsorbed protein (probably albumin) such that it is 'recognized' in some specific way by the macrophages of the various organs. Alternatively, the serum component may be in the nature of an opsonin, the characteristics of which affect the subsequent uptake and distribution of the colloid.

OPSONINS

Disagreement continues as to the requirement for and the nature of opsonizing factors for the phagocytosis of intravenously injected particles. Jenkin and Rowley (1961) claimed that the clearance of both bacteria and inert colloids, such as carbon, was dependent upon the presence of serum opsonins. They induced 'blockade' of the macrophages in mice by giving large doses of intravenous colloidal carbon, following which normal doses of carbon were cleared only slowly from the blood. However, if the carbon suspension was pretreated with foetal pig serum before injection it was cleared at almost the normal rate in blockaded mice. Moreover, the pretreated carbon was cleared more rapidly from the blood of normal (non-blockaded) mice. In contrast, Biozzi, Stiffel, Halpern and Mouton (1960) were unable to demonstrate the existence of opsonizing factors for carbon using similar experimental techniques. Biozzi and Stiffel (1963) also studied the clearance of *Salmonella*, red cells and carbon from the blood of the mouse, and concluded that, although serum opsonins play an important part in the process of phagocytosis of bacteria and red cells, opsonins do not intervene in the phagocytosis of inert particles. Studies by these authors on phagocytosis in colostrum-deprived newborn piglets lacking natural antibodies showed that, in these animals, intravenously injected *Salmonella* are chiefly phagocytosed by spleen macrophages, while the number of bacteria ingested by the Kupffer cells of the liver was small. The lack of bacterial phagocytosis by the Kupffer cells was not due to inefficiency of their phagocytic ability since the cells were still able to engulf large amounts of carbon. The authors concluded that the bacterial uptake by spleen macrophages seemed to be independent of serum opsonins, while the phagocytosis of bacteria by Kupffer cells required previous opsonization. These authors also confirmed the previous observations of Biozzi *et al.* (1963) that pretreatment of carbon particles with normal pig serum or serum from immunized mice did not affect its rate of clearance after intravenous injection. Lirenman, Fish and Good (1967) were also of the opinion that opsonins do not enhance phagocytosis of inert particles in the rabbit, after observing that pre-treatment of colloidal carbon by fresh serum from various species did not prevent the delay in its clearance after blockade produced in rabbits by the prior injection of aggregated bovine serum albumin. The lack of opsonins for inert particles (colloidal gold) in normal rabbit serum was also indicated by the experiments of Filkins, Saba and DiLuzio (1966) using the isolated perfused rat liver as an experimental model. They were, however, able to demonstrate the existence of a nonspecific opsonin in rat serum which facilitated the phagocytosis of colloidal gold. These findings suggest that, in the rabbit, the phagocytosis of inert particles may be independent of serum factors.

If such factors exist, they may be active in certain species only and consequently would not appear to be of universal importance.

Filkins and DiLuzio (1966), using the perfused rat liver, found that carbon uptake in that organ was much reduced when the perfusate comprised plasma previously exposed to fish glue or gelatin (components of the colloidal carbon suspension), when compared to untreated plasma. Murray (1963) reported that blockade occurs only when the surface properties of the blocking and tracer colloids are similar. Thus, when carbon suspended in gelatin was used as a blocking agent, it had no effect on the subsequent clearance of chromic phosphate suspended in saline, but retarded the clearance of chromic phosphate which was suspended in gelatin. In these experiments, different particles stabilized by the same suspending agent, such as gelatin, behaved as identical particles; and identical particles stabilized by different agents behaved as different particles. The apparent contradiction of these findings by the observations of Schapiro, MacIntyre and Schapiro (1966) that blockade produced by the intravenous injection of a massive dose of saccharated iron oxide will appear blocked if tested with subsequent doses of either saccharated iron oxide or chromium phosphate but will clear a test dose of colloidal gold normally, may be related to the fact that these experiments were carried out in rabbits, and Filkins *et al.* (1966) have observed that opsonins for colloidal gold are normally lacking in rabbit serum. However, somewhat similar findings to those of Schapiro, MacIntyre and Schapiro (1966) in the rabbit have been observed by Wagner and Ilio (1964) in the human. The latter authors gave blocking doses of heat-aggregated human serum albumin to normal men and found that the subsequent clearance of tracer doses of the same colloid was much reduced, but there was little effect on the subsequent clearance of colloidal gold or chromium phosphate.

From the above evidence, although apparently contradictory in some cases, it would appear that serum factors may be involved in the phagocytosis of 'inert' particles after intravenous injection. There appears to be considerable doubt as to the specificity of such serum factors, and also doubt as to whether such factors exist for all types of colloids in all species. The findings that factors promoting phagocytosis can be specifically adsorbed from the plasma suggest that these factors may justifiably be called natural antibodies, and, as such, may function during phagocytosis in any of the ways postulated for opsonizing factors discussed in the previous chapter. Several other alternative or additional possibilities also exist. The common observation that blockade produced by one type of colloid does not necessarily result in inhibition of clearance of other types of particulates has been suggested by some as indicating that the specificity of blockade is the result of the existence of different populations of macrophages, each of which is particularly efficient at ingesting certain kinds of particles. Supporting evidence for this concept may be readily observed when liver macrophages containing abundant colloidal carbon may be seen in close proximity to macrophages containing none of this colloid. Moreover, if particles are adsorbed on to 'specific sites' on the surface of the macrophage which are then ingested together with the particle, and the rate of clearance is dependent upon the rate of regeneration of adsorbing sites (Gabrielli and Snell, 1965), then it may be postulated that either each individual macrophage possesses adsorbing sites all of which are more or less specific to one type of colloid, or all macrophages possess a range of adsorbing sites providing a spectrum for limited ingestion of a number of colloids. In either case, the specificity of the adsorbing sites may be related to serum factors and not to the physicochemical characteristics of the colloid concerned. In this latter context, a particle, upon entering the blood stream, immediately adsorbs a layer of protein, probably albumin (Wilkins and Myers, 1966), and the surface properties of the colloid would alter the steric arrangement of the adsorbed protein such that it is 'recognized' in some specific way by the macrophages. A similar process could also be envisaged during which the adsorption of some specific blood component (opsonin) is involved.

As in the case of 'inert' colloidal particles, there can be no general conclusion made as to the requirement for and the nature of opsonizing factors for the phagocytosis of bacteria.

Studies of phagocytosis of bacteria by Kupffer cells in the isolated perfused rat liver (Howard and Wardlaw, 1958; Wardlaw and Howard, 1959) demonstrated that *Staph. aureus*, *B. cereus*, *Corynebacterium murium* and *Str. pyogenes* were phagocytosed better in the absence of serum than in its presence; *E. coli*, *Pseudomonas pyocyanea* and *Proteus mirabilis* were phagocytosed better in the presence of serum and not at all in its absence; and *Str. pneumoniae* was not phagocytosed in the absence or presence of serum. These authors also reported that the opsonic factor for *E. coli* present in human serum was partially removed by heating the serum for 30 minutes at 56 °C, by absorption of the serum with *E. coli*, or by absorption with antigen–antibody complex. Various combinations of these treatments of the serum resulted in complete abolition of the opsonizing factor(s). These findings suggested that, in the case of *E. coli*, naturally-occurring antibody may play an important part in its phagocytosis. Naturally-occurring antibody may also play a part in the phagocytosis of bacteria in experiments performed *in vivo* which show rapid clearance of some organisms in the absence of immune antibodies, as in the case of *Staph. albus* in mice (Cohn, 1962), *S. enteritidis* in mice (Biozzi *et al.* 1960) and *E. coli* in rabbits (Benacerraf, Sebestyen and Schlossman, 1959).

The importance of immune antibodies is demonstrated by the findings of Jenkin and Rowley (1963) that in mice, avirulent strains of *S. typhimurium* were cleared at a much faster rate than the virulent strain. The virulent strain was eliminated from the circulation at a much faster rate after pretreatment (opsonization) with normal sera from different species of animals, and pig serum was particularly effective in this respect. These authors also observed that natural antibodies in heterologous sera may be as effective as immune antibodies in promoting phagocytosis of virulent *S. typhimurium*. Spiegelberg, Miescher and Benacerraf (1963) have also shown that *E. coli* treated with immune antibody were cleared more rapidly than untreated *E. coli*. In both cases, mice 'decomplemented' by the injection of aggregated human γ-globulin exhibited reduced rates of clearance of *E. coli*, although, in the relative absence of complement, the *E. coli* treated with immune antibody were still cleared more rapidly than the untreated organisms. The rate of clearance of *E. coli* in decomplemented mice was restored to normal by injections of fresh mouse serum but not by injections of heated normal mouse serum. These findings indicate that complement has an important effect in promoting phagocytosis but its requirement is not essential, as is exemplified by the continuing phagocytosis in 'decomplemented' animals. Moreover, the findings suggest that immune antibody may promote phagocytosis in the relative absence of complement. In the interpretation of these results, it must be remembered that 'decomplementation' was not absolute, and that the haemolytic complement levels in the sera was still in the order of 5 per cent of the original activity.

Stiffel *et al.* (1964) examined the rate of clearance of intravenously injected *S. typhimurium* labelled with ^{131}I in normal mice and in mice genetically lacking haemolytic complement activity, and found that complement-lacking mice had a normal ability to phagocytose this organism. Decomplementation *in vivo* with unrelated antigen–antibody complexes did not affect the rate of clearance, but markedly reduced the effectiveness of specific rabbit antibody in opsonizing *S. typhimurium* for phagocytosis *in vitro*. The results indicated that, in the case of *S. typhimurium*, neither bacterial phagocytosis nor the effect of immune opsonins are related to the serum complement titre measured in terms of haemolytic activity.

From the findings quoted above, and from the numerous other studies performed on this topic, it may be concluded that there can be no general conclusions made as to the requirement for and the nature of opsonizing factors for the phagocytosis of circulating bacteria. It would appear that the precise requirements of such factors may vary according to the micro-organism concerned. In most cases, it seems that phagocytosis is promoted by, but may not be absolutely dependent on, the presence of natural or immune antibodies and complement. Whether all or some components of complement are important in this respect is unknown. It is possible that serum opsonins play a more important role in promoting the intracellular killing of bacteria after they have been phagocytosed by macrophages, rather than in events preceding phagocytosis.

BLOCKADE

The term 'blockade' is used to describe the situation when, following the intravenous injection of a colloid which has been taken up by the macrophages, the clearance of subsequent doses of the same or a different colloid is significantly delayed. Blockade has been attributed to the satiation of the phagocytic cells, to the saturation of the phagocytic mechanism, or to the depletion of serum factors necessary for phagocytosis consequent upon the uptake of the first injection of colloid.

It has been observed that intravenously injected colloids are removed from the circulation in an approximately exponential fashion, but, in contradiction to the exponential concept, the rate of removal of colloid from the blood proceeds at a constant rate instead of being proportional to the amount present. This contradiction has been explained by Benacerraf *et al.* (1957) as being due to the opposite effects of two phenomena – the saturating effect of phagocytosed carbon, which decreases the efficiency of clearance progressively throughout the experiment, being counterbalanced by the increased efficiency with which macrophages can extract particles from the circulation as the blood concentration decreases. This invokes the concept of a constant phagocytic rate which yields linear instead of exponential kinetics, and the concept of a gradually progressive blockade. The concept of a progressive satiation of the phagocytic cells was made questionable by the findings of Parker and Finney (1960) who reported that a second injection of carbon given 1–3 hours after the first did not show a slowing of the rate of clearance, and that 'blockade' only occurred after a latent period of about $4\frac{1}{2}$ hours. This behaviour does not fit the concept of simple satiation of macrophages as being an important factor in producing blockade.

More recently, emphasis has been placed on the possibility that blockade may be principally due to the depletion of essential serum opsonins, either naturally-occurring or the result of immunization, during the phagocytosis of particles consequent upon the first injection of colloid (Jenkin and Rowley, 1961; Murray, 1963; Normann and Benditt, 1965). Thus it is suggested that blockade is due to the lack of available opsonins for the particles. The evidence for the existence and importance of serum opsonins has been examined in the previous section of this chapter. The proven existence of specific opsonins also provides an acceptable explanation for the common observation that blockade produced by one type of colloid does not necessarily result in the inhibition of clearance of other types of particulates. However, this mechanism for the induction of blockade, like the satiation of phagocytic cells, does not adequately explain the latent period in the production of blockade observed by Parker and Finney (1960) nor is it in agreement with the interesting observations of Dobson, Kelly and Finney (1967). The latter authors found that four repeated intravenous injections of colloidal carbon in mice given at 1–2 hourly intervals *prolonged* the latent period before blockade was produced and caused an *increase* in the rate of removal of the colloid which was very marked when the dose of colloid was increased. This effect, the antithesis of blockade, was evidenced by a rate of removal many times as great as the rate of removal of the first injection. The phenomenon occurred too rapidly to be the result of cellular proliferation, and seemed to be too rapid to be the result of synthesis of new opsonins or other serum factors. The fact that neither actinomycin D nor puromycin treatment had any inhibitory effect on the stimulation of phagocytosis by repeated particle injection suggests that the observed increase in phagocytic activity was not due to the elaboration of new protein for cell membrane or adaptive enzyme synthesis.

Blockade produced by one type of particle does not necessarily result in the inhibition of clearance of a different type of particle. This has been suggested by some as indicating that the specificity of blockade is the result of the existence of clones of macrophages, each clone having a proficiency for ingesting a narrow range of particles. Evidence in support of this concept obtained by the light microscopical observation of Kupffer cells containing abundant carbon particles adjacent to cells containing little or no visible carbon is not supported by the electron microscopical finding that, in these cases, all Kupffer cells contain at least some carbon particles (Wiener, 1967). However, the latter observations do not eliminate the possibility that some macrophages have a greater potential capacity for particular types of particles but may be 'recruited' to take up other types of particles when the specialized population are becoming satiated. Electron microscopical studies have also shown that, in blockaded animals, very few carbon particles are in contact with the surface membranes of Kupffer cells; and this contrasts with control animals where the carbon particles are found either on the cell surface membrane or within cytoplasmic vacuoles of the Kupffer cells (Wiener, 1967). This suggests that the failure to endocytose carbon during blockade is the result, at least in part, of a defect in the surface attachment phase of phagocytosis. Thus, the concept of the depletion or inhibition of opsonins, calcium ions and other factors which affect the attachment of particles to the cell membrane playing some part in blockade receives support from these observations. An additional explanation is afforded by the concept of particles being adsorbed on to 'specific sites' on the surface of the macrophage which are ingested together with the particle, and the rate of clearance being dependent upon the rate of regeneration of such adsorbing sites (Gabrielli and Snell, 1965). In this case, the degree and duration of blockade would be dependent upon the rate of regeneration of the adsorbing sites and the specificity of blockade would be the result of the adsorbing sites having a limited specificity related to their own configuration, to the specificity of opsonic factors for particular particles, or to the presence of cytophilic antibody on the cell membrane. Another mechanism involved in the escape from blockade has been suggested as being due to the proliferation of a new Kupffer cell population, and it has been demonstrated that recovery from blockade is accompanied by Kupffer cell proliferation, and that such cells are able to divide despite the fact that they contain ingested colloidal material (Kelly *et al.* 1962).

THE QUANTITATIVE ASSESSMENT OF PHAGOCYTOSIS *IN VITRO*

Most studies involving the quantitative assessment of the phagocytic activity of macrophages have been carried out *in vivo* using one of the clearance techniques outlined in the preceding part of this chapter. While these techniques appear to give a true quantitative assessment, they only measure the phagocytic activity of those macrophages which are in contact with the circulating blood – primarily the Kupffer cells of the liver – and do not necessarily reflect the phagocytic potential of the very large numbers of macrophages which are not in contact with the circulating blood and which do not have ready access to intravenously injected circulating particulates. Studies using the isolated perfused liver can provide very useful information on various factors affecting phagocytosis, but again only reflect the phagocytic activity of Kupffer cells. As yet, there are no techniques available for the quantitative assessment of all the macrophages of the body at one and the same time, although vital staining can provide useful qualitative information in this respect. The advantages of being able to measure the phagocytic activity of macrophages *in vitro* are manifold, but extrapolation of *in vitro* results to events which may occur *in vivo* must be carried out with caution. However, *in vitro* techniques enable the phagocytic activity of various populations of macrophages, harvested from various sites (blood, peritoneum, alveoli, inflammatory exudates) to be assessed, and also allow a more controlled study of such factors as opsonization, and morphological and metabolic changes during phagocytosis.

Macrophages for *in vitro* studies may be obtained by washing out the unstimulated or 'stimulated' peritoneal cavity with some suitable solution (physiological saline, Hank's BSS, tissue culture medium), from the alveoli by flushing out the lungs via the trachea, from inflammatory exudates, from the buffy coat of the blood, and by mincing various organs and separating the macrophages from other cells either by filtration through glass columns or in an electromagnetic field after the macrophages have selectively ingested iron particles. Whatever the source, and peritoneal macrophages have been most widely used in such studies, the general principles involved in the quantitative assessment of phagocytosis *in vitro* are that a known number of bacteria or inert particles are exposed to a known number of macrophages in a suitable culture medium. The techniques used for viable bacteria and inert particles have certain fundamental differences.

BACTERIA

Most *in vitro* studies use a technique similar to that which was originally described by Jenkin and Benacerraf (1960). The general principles involved are that a known number of harvested macrophages are allowed to settle and adhere to glass coverslips in flasks or to the walls of Leighton tubes, and known numbers of bacteria are added to the macrophage cultures, so that the ratio of bacteria to cells is in the order of one bacterium to between 100 and 500 macrophages (Jenkin and Benacerraf, 1960; Jenkin, 1963). Flasks or tubes containing no macrophages act as controls for assessing bacterial growth; or, better still, bacteria are added to suspensions of macrophages killed by heating at 100°C for 15 minutes. All cultures are incubated at 37°C, and at intervals thereafter, known volumes of supernatants from test and control cultures are plated on suitable agar medium (depending on the organism used) to determine the number of bacteria in the supernatants. The macrophages are washed and then disrupted by mechanical means, and the resulting macrophage suspensions are plated out on to a suitable medium to ascertain the number of viable organisms phagocytosed by the cells. For each time interval it is possible to assess the extent of phagocytosis by the cells and the survival of organisms within the cells by the following calculation. If

N_1 = number of organisms in supernatant of control cultures,
N_2 = number of organisms in supernatant of test cultures,
C = number of organisms in cells from test cultures,

then $(N_1 - N_2)/N_1 \times 100$ = percentage of organisms phagocytosed by the cells, and $C/(N_1 - N_2) \times 100$ = percentage of ingested organisms surviving within the cells. The results can then be plotted as log per cent survival or log per cent phagocytosis against time.

The above technique is very useful for assessing the *in vitro* effects of factors affecting the bacteria (virulent and non-virulent strain differences, opsonins, phage, etc.) and the macrophages (source, stimulant and depressant agents, metabolic changes, etc.). A simpler assessment of the phagocytic rate may be obtained by microscopy by counting the number of cells containing bacteria and scoring them against those which do not contain organisms; and this also enables the mean number of bacteria per phagocytic cell to be determined.

INERT PARTICLES

Vaughan (1965 *a*) developed an ingenious apparatus which enabled known numbers of phagocytic cells and inert particles to be deposited together on a millipore cellulose ester filter membrane. After particles (erythrocytes, cellulose powder, zymosan granules) and cells had been incubated in suitable medium within the apparatus for 45 minutes, the cells were fixed and stained. The membrane, made translucent by mounting in a suitable medium, was examined under the microscope and the percentage of cells containing one or more particles was recorded.

Perkins, Nettesheim, Morita and Walburg (1967) assessed the phagocytic activity of macrophages by incubating them with opsonized sheep red cells (SRBC). At intervals, aliquots of the suspensions were exposed to a short hypotonic 'shock' which resulted in complete lysis of extracellular SRBC, while the macrophages and engulfed SRBC were undamaged. The numbers of extracellular SRBC were assessed from the haemoglobin concentration in the lysate.

EXOGENOUS AND ENDOGENOUS FACTORS AFFECTING PHAGOCYTOSIS

AGEING

There have been limited studies on phagocytic activity in embryos. Jenkin and his colleagues (Karthigasu and Jenkin, 1963; Reade and Jenkin, 1965) studied the uptake of colloids and bacteria in chicks and rats and found that, although the capacity for phagocytosis increased with age, the macrophages of the foetal chick and rat are less capable of destroying endocytosed bacteria than those of the adult animals. The differences in intracellular killing are probably related to the development of serum opsonins. Benacerraf *et al.* (1957) used the carbon clearance technique to examine the phagocytic activity in rats of various ages, and found that the phagocytic index K increases slowly from birth up to 3 weeks of age, and then decreases in the adult and older animal. When the corrected phagocytic index α, relating K to the weights of liver and spleen, was determined, it was found that macrophage activity is highest 10 days after birth and decreases progressively thereafter as the animal becomes older. In relation to total body weight, the liver and spleen are small at birth but are endowed with greater phagocytic activity (possibly due to higher blood flow); and the liver and spleen increase in relative weight in the adult rat and decrease again in the older animal.

There have been a limited number of studies on the effect of age upon phagocytic activity in man. Antonini, Cappelli, Citi and Serio (1964) studied the clearance of intravenously injected ^{198}Au in 26 healthy subjects in the age range 20–84 years, and observed that there was a progressive fall in the index of phagocytic activity K as age increased. Although these authors selected subjects within a narrow body weight range (62–71 kg), they could not exclude the possibility that the relative reduction in liver weight as age progresses may be accompanied by a numerical decrease in the number of Kupffer cells. The authors were also aware that hepatic blood flow is reduced in older persons, and this could account for the results which they obtained. They attempted to investigate this aspect by catheterizing the hepatic veins, but stopped the investigation on finding that the procedure caused atrial fibrillation in older subjects. A progressive reduction in phagocytic activity during senescence in man has also been observed by Wagner, Migita and Solomon (1966) who measured the clearance of heat-aggregated human serum albumin and radioactive colloidal gold.

HORMONES

It was shown by Nicol (1935) that oestrogenic hormones stimulate the appearance and activity of macrophages in the endometrium of the uterine horns of the guinea-pig, as shown by their capacity to take up and concentrate trypan blue. He noted that their number in the endometrium varied at different stages of the oestrous cycle and was greatest at the time of maximum oestrogen concentration in the blood. Macrophages were rarely seen after ovariectomy but reappeared following oestrogen injection. From his findings, Nicol deduced that the appearance and activity of the macrophages in the endometrium were under the control of ovarian oestrogen. Using the carbon clearance technique, Nicol and Vernon-Roberts (1965) showed that two peaks of phagocytic activity by the sinusoidal macrophages occur during the oestrous cycle in the rat and the mouse, one during the follicular phase (at pro-oestrus) and the other during the luteal phase (at metoestrus) at the time of endometrial degeneration, which were accompanied by a more prolonged invasion by macrophages into the endometrium. After ovariectomy, the level of phagocytic activity fell and, although subject to fluctuations, did not attain the levels

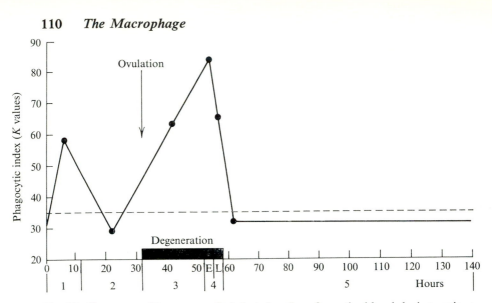

Fig. 32. Clearance of intravenously injected carbon from the blood during various stages of the oestrous cycle in the rat. (1) Pro-oestrus; (2) early oestrus; (3) late oestrus; (4) E, early metoestrus; L, late metoestrus; (5) dioestrus. ●———● Normal animals; ----- ovariectomized controls. Reproduced with permission from Nicol *et al.* (1964).

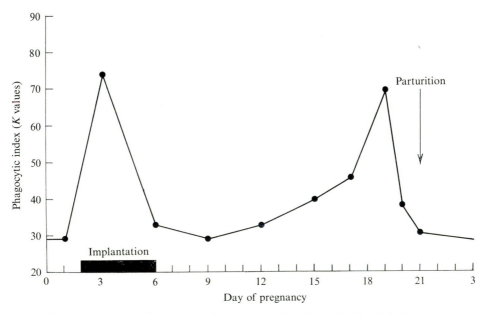

Fig. 33. Clearance of intravenously injected carbon from the blood during pregnancy and immediately *post partum* in the rat. Reproduced with permission from Nicol *et al.* (1964).

seen prior to the removal of the ovaries. Two peaks of increased phagocytic activity were also observed during pregnancy, one at the time of implantation and the other towards the end of pregnancy before parturition, and these were also accompanied by a more prolonged invasion by macrophages into the decidua (Figs. 32, 33). The peaks of increased phagocytic activity observed during the reproductive cycle during pregnancy corresponded with those periods when oestrogen levels in the blood are increased (Nicol and Vernon-Roberts, 1965).

The above observations that phagocytosis normally appeared to be mainly under the influence of blood oestrogen levels, and the finding that the administration of natural and synthetic oestrogens resulted in marked stimulation of phagocytosis (Nicol *et al.* 1964), led Nicol, Vernon-Roberts and Quantock (1965*b*) to investigate the influence on phagocytic activity of various hormones, both when acting alone and when acting together with oestrogen. Their results may be summarised as follows: (1) the three principal naturally-occurring oestrogens, oestradiol-17β, oestrone and oestriol, are strong stimulants of phagocytosis. (2) Testosterone has no effect on phagocytosis when administered alone, but doses which inhibit the oestrogenic response of the reproductive tract to oestrone and oestriol also inhibit the stimulant effect of these oestrogens on phagocytosis. In contrast, testosterone markedly potentiates the stimulant effect of oestradiol-17β on phagocytosis even in doses at which the oestrogenic response is inhibited. (3) Progesterone is a mild stimulant of phagocytosis. Doses which inhibit the oestrogenic response of the reproductive tract do not affect oestrogen-produced stimulation of phagocytosis. (4) Small doses of cortisone stimulate phagocytic activity, but high doses depress phagocytosis and oestrogen-produced stimulation of phagocytosis. (5) Thyroxine has no effect on phagocytosis when acting alone, but markedly potentiates oestrogen-produced stimulation of phagocytosis. (6) The hormones of the anterior lobe of the pituitary are all stimulants of phagocytosis; and, with the exception of growth hormone and follicle-stimulating hormone, they act in a synergistic manner with oestrogen. (7) The hormones of the posterior lobe of the pituitary have no effect on phagocytosis. It was concluded that the endocrine system exerts a homeostatic influence on macrophage phagocytic activity, and it was suggested that oestradiol-17β is the principal oestrogen concerned in the control of macrophage function.

Using clearance techniques, it has been shown that, in general, natural and synthetic steroid and non-steroid oestrogens possessing high oestrogenic potency are strong stimulants of phagocytosis (Nicol *et al.* 1964). However, α,α-dimethyl-β-ethyl-allenolic acid, which has a very high oestrogenic potency, does not stimulate the macrophages (Biozzi *et al.* 1957), and 1,3,5,estratriene-3,16β-diol has the same stimulant effect as does oestriol (Heller, Meier, Zucker and Mast, 1957) but has only approximately one-hundredth of its oestrogenic potency. Similarly, diethylstilboestrol and diethylstilboestrol dimethyl ether are equally potent stimulants of phagocytosis, yet the oestrogenic potency of the dimethyl ether is only one-twentieth of that of diethylstilboestrol. In contrast, 2:3-di(*p*-hydroxyphenyl)2-hexene has an oestrogenic potency equal to that of diethylstilboestrol but is a much weaker stimulant. From these findings, it would appear that the macrophage-stimulating effect does not necessarily run in parallel with oestrogenic potency (Nicol *et al.* 1964). To investigate this relationship further, Nicol, Vernon-Roberts and Quantock (1966*a*, *b*, *c*, 1967*b*) investigated the effects of various combinations of oestrogens and anti-oestrogens on the phagocytic activity of macrophages and on the reproductive tract in intact and gonadectomized male and female mice. They showed that doses of testosterone, progesterone and 17α-ethynyl-19-nortestosterone which inhibited the response of the reproductive tract to oestradiol-17β, do not inhibit the stimulant effect of oestradiol on phagocytosis. They also showed that oestradiol, 16-epioestriol, 2-methoxyoestrone and 2-hydroxyoestradiol-17β, which are metabolites of the principal naturally-occurring oestrogen, oestradiol-17β, are oestrogenic and stimulate macrophage activity; when each of these metabolites was administered together with oestradiol-17β, there was mutual inhibition of the effects of both oestrogens on the reproductive tract but no inhibition of their stimulant effects on phagocytosis. In contrast, oestriol and 17α-ethynyl-19-nortestosterone inhibited the stimulant effect of diethylstilboestrol on phagocytosis; and MER-25 inhibited the stimulant effects of both oestradiol-17β and diethylstilboestrol. It was concluded that these findings clearly demonstrated that the action of oestrogen on the macrophage system is independent of its action on the reproductive tract although the two biological activities are contained within the same molecule. Nicol and his colleagues also examined their results in relation to the structure of the various compounds used in their experiments, and suggested that steroid and nonsteroid oestrogens compete with anti-oestrogens for attachment to a common receptor site; and the differences in the effects of interaction on the macrophage system and on the reproductive tract may be the result of selective differences in the binding effect to proteins in the two target areas.

The mechanisms whereby oestrogens act on macrophages are unknown and can only be postulated from available evidence (reviewed by Vernon-Roberts, 1969*b*). It appears that oestrogen may have physiological and pharmacological effects on the macrophage system depending on the magnitude of dose used, and it is possible that they are mediated by different mechanisms. Marked proliferation of liver macrophages has been observed on the second and third days after oestrogen injection (Kelly, Dobson, Finney and Hirsch, 1960), and this coincides with the time taken to reach maximum phagocytic activity (Heller *et al.* 1957; Kelly *et al.* 1960). Accumulated evidence also suggests that oestrogens may increase macrophage cell membrane permeability as a result of binding of the oestrogen molecules to protein in the cell membrane, and this concept has received support by light and electron microscopical studies on the effects of oestrogen on thorotrast uptake in the liver (Grampa, 1967).

It has been widely shown that cortisone will reduce the resistance of animals to many kinds of infections. However, it has also been observed that cortisone, under certain conditions, will enhance the resistance of animals. This bimodal effect may be related to the depressant effect of large doses of cortisone on the phagocytic activity of macrophages, and the stimulating effect of smaller doses. Doses of cortisone in excess of 10 mg per kg body weight cause marked depression of phagocytic activity in guinea-pigs, rats and mice (Nicol *et al.* 1964, 1965, 1967*a*; Benacerraf *et al.* 1957; Lurie, 1960; Snell, 1960*a*, *b*.) In contrast, doses of cortisone below 5 mg per kg stimulate phagocytosis (Lurie, Zappasodi, Dannenberg and Schwartz, 1951; Snell, 1960*a*, *b*; Nicol *et al.* 1965, 1967*a*). Snell (1960*a*, *b*) reported that, like cortisone, cortisol (hydrocortisone) at dose levels below 5 mg per kg stimulated phagocytic activity, but depressed phagocytosis as the dose was increased above this level. Corticosterone also produced this effect to a lesser extent. The effect seemed relatively specific to these three compounds since 11-deoxycortisol (Reichstein's Compound S), 11-deoxycorticosterone (DOC) and thirty-six other steroids tested had no such effect. All the steroids used eventually caused depression of phagocytosis when administered in high doses. Nicol *et al.* (1967*a*) showed that cortisol and prednisone are more potent than cortisone in depressing phagocytosis. In contrast to Snell's findings they reported that DOC stimulated phagocytosis even when the dose was increased to 500 mg per kg.

Lurie *et al.* (1951) observed that rabbits treated with cortisone (1 mg per kg) showed greater phagocytic activity than the untreated controls. Following the administration, by inhalation, of human-strain tubercle bacilli, it was found that, although phagocytosis of the tubercle bacilli was greater in the treated rabbits, cortisone treatment interfered appreciably with the capacity of the macrophages to digest the endocytosed bacilli. Cortisone also deprived the macrophages of their innate capacity to inhibit the multiplication of the bacilli within their cytoplasm. Lurie (1960) later showed that cortisone also retards the transport of tubercle bacilli from the site of invasion. It appears that the increased intracellular survival of phagocytosed tubercle bacilli after cortisone treatment reported by Lurie is confined to this particular organism since, also in the rabbit, Robinson (1953) and Crabbé (1956) have reported that small doses of cortisone enhance the phagocytosis of other types of bacteria by macrophages and increase protection against infection. These observations on the effects of low doses of cortisone in animals are in marked contrast to the well-recognized dangers of reduced resistance to infection resulting from the doses of corticosteroids commonly used in various forms of therapy in the human at the present time.

It appears that corticosteroids may depress the phagocytic activity of macrophages in several ways. It has been reported that cortisone stabilizes the cell membrane of the macrophage and brings about an inactive state (Eyring and Dougherty, 1955), and that macrophages exposed to cortisol *in vitro* become very quiescent (Dougherty, 1961). These observations are in agreement with the electron microscopical findings of Wiener (1967) that cortisone treatment inhibits the attachment of carbon particles to Kupffer cells. Cortisone also stabilizes the lysosomal membranes of phagocytic cells (Weissmann and Thomas, 1963) and inhibits the release of hydrolytic enzymes after the endocytosis of materials (Donaldson, Marais, Gyi and Perkins, 1956).

LIPIDS

Certain lipids possess the capacity to modify the phagocytic activity of macrophages following intravenous injection of particulate suspensions. Stuart and Cooper (1962) made a detailed analysis of the effects of glyceryl trioleate (triolein) and ethyl stearate in mice. Triolein was found to enhance the rate of removal of particulate material from the blood stream, while ethyl stearate caused a significant depression of this activity; these substances seemed to affect directly the macrophages of the liver and spleen, and there were no detectable histological changes in other cells or organs. Further studies showed that the glyceryl esters of fatty acids (glyceryl trioleate, glyceryl tricaprate) stimulate phagocytic activity, whereas the alkyl esters (ethyl stearate, ethyl oleate, butyl oleate, ethyl palmitate) and cholesterol esters (cholesterol oleate) depress phagocytic activity (Stuart, 1963). Histological studies during these investigations showed that stimulation by glyceryl esters appeared to be due to a direct effect on Kupffer cells thereby enhancing their phagocytic efficiency. The depression of phagocytosis by alkyl and cholesterol esters of fatty acids may be related to direct effects causing the death of macrophages in the liver and spleen, to decreased numbers of phagocytic cells, to necrosis of the spleen, and to a blockading effect of the endocytosed lipid particles.

Electron microscopical and autoradiographic studies of peritoneal macrophages stimulated *in vitro* by incubation with glyceryl trioleate (Carr and Williams, 1967) support the concept of a direct effect on macrophages by demonstrating the presence of tritiated glyceryl trioleate in large amounts near the surface of macrophages and within them. These studies also showed that the cell processes become longer and more prominent after triolein stimulation. *In vitro* studies by Cooper (1964) also indicated that contact between the triglyceride particles and macrophages is necessary for the stimulation of phagocytosis; and, by examining the effect of a large number of lipids, Cooper suggested that the chemical structure of the fatty acid radical determines the degree of activity of the triglyceride.

The knowledge that crude extracts of lipid-containing fractions of cell membranes from various sources had been shown to affect the phagocytic activity of macrophages stimulated Bolis, Kessel and Petti (1967) to investigate the effects of cholesterol, lecithin, α-lysolecithin and phosphatide–peptide fraction (isolated from rat liver cell membrane) on carbon clearance, on the *in vitro* macrophage permeability to streptomycin (as assessed by the intracellular killing of *Brucella abortus*), and on the release of hydrolytic enzymes from isolated lysosomes *in vitro*. They found that lecithin, lysolecithin and phosphatide–peptide stimulated carbon clearance and promoted the release of enzymes from isolated lysosomes, but cholesterol had no effect on either parameter. None of the compounds used has any detectable effect on the permeability of macrophages to streptomycin.

MICRO-ORGANISMS

Variations in phagocytic activity, assessed by clearance techniques, have been observed in experimental animals and man during infections with various micro-organisms. Wagner, Ilio and Hornick (1963) studied the clearance of ^{131}I-labelled aggregated human serum albumin in patients suffering from infections, and found that the maximum rate of phagocytosis was increased in pneumococcal pneumonia, typhoid fever and pneumonic tularaemia but was reduced during infection with sandfly fever. The enhanced rates of phagocytosis were observed after the patients showed clinical signs of recovery from infection, and it was postulated that enhanced phagocytosis may be related to the 'scavenging' activity of macrophages to debris resulting from the infection.

In experimental animals, infection with living organisms causes a rapid and prolonged increase in phagocytic activity. Biozzi *et al.* (1957) showed that infection of mice with virulent strains of *Salmonella* and *Mycobacteria* produced a marked stimulation of phagocytic activity within 24 hours of inoculation which was accompanied by a rapid increase in the size of the liver and spleen; the phase of stimulation was followed by a marked and rapid drop in phagocytic activity as the animals became clinically ill and succumbed to the infection. Infection with non-virulent strains also caused a rapid increase in phagocytic activity, but this was followed by a progressive return to normal levels as the animals recovered. Böhme (1960) investigated carbon clearance in genetically resistant and susceptible strains of mice infected with *S. typhimurium* and obtained some unexpected results. In susceptible animals, phagocytic activity had risen significantly one day after infection, reached maximum levels on the fourth day, and dropped rapidly to control level on the sixth day. The animals succumbed about eight days after infection. In contrast, the resistant strain of mice did not exhibit any rise in phagocytic activity until the eighth day after infection, reached a maximum about the fifteenth day, and returned slowly to normal levels at about thirty days when the experiment was terminated and the animals had recovered. At the time of active infection, microscopy revealed that the splenic lesions characteristic of this infection were more numerous and more destructive in the susceptible animals. It was concluded that the most likely explanation of the findings is that, for some unexplained reason, the susceptible mice have an innate susceptibility not related to their phagocytic ability since, prior to infection, both strains exhibit identical phagocytic activity and ability to form agglutinins against gram-negative bacteria.

The lipopolysaccharide endotoxins of gram-negative bacteria are structural components of the bacteria and are released on death and dissolution of the organisms. Endotoxins are also referred to as lipopoly-saccharide antigens, glucolipid antigens, somatic antigens, bacterial pyrogens, Schwartzman-active toxins, tumour-necrotizing toxins or shock-producing toxins. After intravenous injection they produce fever, varying degrees of shock leading to vasoparalysis and stagnation of blood flow, metabolic disturbances (hyperglycaemia, elevation of blood lactate and pyruvate, depletion of liver glycogen), immediate leucopaenia, and mural thromboses in the heart and small vessels. If two intravenous injections are given less than 24 hours apart, the second injection is rapidly followed by a systemic reaction, designated the generalized Schwartz-man reaction, accompanied by cortical necrosis in the kidneys due to fibrinoid deposition in blood vessels. When repeated daily injections are given for four or five days the recipient animal acquires a state of tolerance and toxic effects do not appear after subsequent injections. However, if the tolerant animal is given a large blockading dose of an inert colloid intra-venously, tolerance is lost. Moreover, if colloids and endotoxin are administered intravenously together, the toxic effects of small doses of endotoxin are very much enhanced. Endotoxins which circulate in particulate form are cleared from the circulation by the macrophages of the liver and spleen (Thomas, 1957; Freedman, 1960), and in some way yet unknown, the endotoxins are then detoxified by these cells. The Schwartz-man phenomenon which follows two doses of endotoxin given within 24 hours, may be related to the observation that the phagocytic activity of liver and spleen macrophages is rapidly and markedly depressed for about 24 hours after a single injection of endotoxin (Benacerraf and Sebestyen, 1957), and it may be assumed that their ability to handle endotoxin is also depressed during this period. It has also been shown that, 48 hours after a single injection of endotoxin, phagocytic activity is elevated and remains so for at least six weeks despite further injections of endotoxin (Benacerraf and Sebestyen, 1957), and this would appear to account for the appearance of tolerance; and also explains the loss of established tolerance after the injection of blockading doses of colloids. Presumably, inhibition of phagocytic activity interferes with the removal of circulating endotoxins from the blood by the macrophages, and allows them to exert their toxic effects on other tissues.

It has been shown that *Mycobacteria* and some of their constituents are potent stimulants of phagocytic activity. Living and killed virulent, avirulent and BCG strains of *Myco. tuberculosis*, *Myco. phlei*, *Myco. fortuitum* and the killed *Myco. butyricum* present in Freund's complete adjuvant are all effective stimulants (Biozzi *et al.* 1957; Böhme, 1960; Böhme and Bouvier, 1960; DiCarlo *et al.* 1963; Nicol, Quantock and Vernon-Roberts, 1966). Various extracted fractions of killed mycobacteria also stimulate phagocytosis for a prolonged period after a single injection (Böhme and Bouvier, 1960; Biozzi, Halpern and Stiffel, 1963), but this is preceded by a fall in phagocytic activity which lasts for some hours (Böhme and Bouvier, 1960). Thus these extracts of mycobacteria simulate the action of endotoxins in the pattern of their effects on phagocytosis. Moreover, the prior injection of mice with *Salmonella* endotoxin protects them against subsequent infection with *Myco. fortuitum* (Böhme, 1960).

Extracts from the cell walls of yeasts, which consist of lipids, polysaccharides and proteins, have been shown to stimulate macrophage phagocytic activity. Crude extracts consisting of a mixture of the constituents ('zymosans'), a purified polysaccharide constituent ('glucan'), and a polysaccharide-free extract comprising lipids alone ('restim'), have all been claimed to contain the active components. It would appear that part of the mechanism of action, like that of oestrogen, is the stimulation of proliferation in liver macrophages (Kelly *et al.* 1962).

IRRADIATION

It is well established that irradiation can impair specific immune responses and increases the susceptibility of animals to infections. It is also recognized that macrophages are particularly resistant to damage by X-rays. After whole-body irradiation, macrophages of lymphoid and haemopoietic tissues are loaded with the debris of cells which have been damaged by X-rays, and kiloroentgen doses of X-rays do not inhibit the phagocytic activity of peritoneal macrophages *in vitro* (Perkins, Nettesheim and Morita, 1966) or of macrophages in organ cultures of lymph nodes and thymus (Gilman and Trowell, 1965).

The effect of irradiation on the clearance of intravenously injected particulates *in vivo* has been studied by a number of workers, and the results vary considerably; stimulation, depression, and no change in clearance rates have all been reported. These results have been summarized by Šljivić (1970*a*), who also investigated the rate of clearance of colloidal carbon in mice up to 7 days after whole-body irradiation with supralethal doses of X-rays and up to 8 weeks after exposure to mid-lethal and sub-lethal doses. During the first few days after irradiation there was little or no change in carbon clearance rates, but thereafter the clearance rate was increased up to two-fold in irradiated animals and there was a clear dose–response relationship. Further studies (Šljivić, 1970*b*) showed that treatment of mice with bone marrow or lymph node cells, or with anti-biotics, partially reversed the stimulating effect of irradiation on phago-cytic activity. Irradiation of the rear part of the body, including the whole abdomen, resulted in a faster clearance of carbon, but irradiation of the front half of the body failed to do so. Phagocytic activity was not altered by irradiation in germ-free mice, but was markedly enhanced in irradiated conventional and specific pathogen-free mice. Administration of anti-biotics to irradiated conventional animals resulted in a partial reversal of the effects of irradiation on phagocytosis. It was concluded that the enhancement of phagocytosis after irradiation was probably not due to a direct effect on the macrophages but was the result of the facilitation of bacterial invasion through damage to the intestinal mucosa, and that the liver macrophages were stimulated in the process of ingesting the cir-culating bacteria.

Despite the apparent absence of an inhibitory effect on phagocytosis, it appears that irradiation may impair the ability of macrophages to digest phagocytosed bacteria. Donaldson *et al.* (1956) found that irradiation suppressed the ability of macrophages to digest chicken erythrocytes and *Candida in vivo* and *in vitro* although phagocytosis was unchanged, and Gordon, Cooper and Miller (1955) observed that *K. pneumoniae* were cleared normally from the blood stream in irradiated rabbits, but the organisms subsequently reappeared in the blood stream, and the animals died.

MISCELLANEOUS FACTORS

It has been generally reported that thymectomy in newborn and adult animals, bursectomy with and without thymectomy in birds, and splenectomy in various species has little or no effect on macrophage activity (Biozzi *et al.* 1960; Schooley *et al.* 1965; Cooper *et al.* 1966). However, there is elevation of phagocytic activity during the runting syndrome after thymectomy in mice (Schooley *et al.* 1965). The proliferative stage of the graft-versus-host (GVH) reaction is characterized by hyperphagocytic activity of the macrophage system (Howard, 1961*a*), which probably represents a host reaction to the accumulation of cell breakdown products. The enhanced phagocytic response (and mortality) during the GVH reaction is greatly reduced by splenectomy 2–8 days, but not 11 days, after the injection of parental-strain spleen cells in adult F_1 hybrid mice (Biozzi, Howard, Stiffel and Mouton, 1964). These findings may suggest that splenectomy removes a large focus of the GVH reaction since many of the donor cells probably settle in this organ, and thus also removes the source of much of the cell debris.

Recent studies have shown that anti-macrophage serum (AMS) and anti-lymphocyte serum (ALS) can depress the phagocytic activity of macrophages. Loewi, Temple, Nind and Axelrad (1969) reported that the presence of 20 per cent anti-guinea-pig macrophage serum in cultures of washed peritoneal macrophages caused lysis and agglutination of up to 90 per cent of the cells within one hour. The presence of 2 per cent AMS caused agglutination without lysis, and lysis of very few cells was seen when ALS was substituted for AMS. The intravenous injection of AMS or ALS depressed carbon clearance, but AMS alone failed to affect the immune response of guinea-pigs to sheep red cells. Chare and Boak (1970) have reported that a single injection of horse anti-mouse lymphocyte serum, rabbit anti-mouse lymphocyte serum, normal horse serum (NHS), normal rabbit serum (NRS) or syngeneic dead cells into mice resulted in a depression of phagocytic activity measured 2–8 hours after the injection. This depression is not a specific or direct property of ALS since NHS, NRS and syngeneic dead cells also had the same effect. Increased or decreased phagocytic activity was observed 24 hours after ALS injection by Chare and Boak, but Sheagren, Barth, Edelin and Malmgren (1969) reported that the intravenous injection of ALS produces immediate, profound and prolonged blockade of phagocytic activity as measured by the clearance of carbon or aggregated human serum albumin: ALS administered by the intravenous route was more effective than when administered intraperitoneally.

The effects upon phagocytosis of a large number of related and unrelated factors, too numerous to be usefully chronicled here, are encountered in the literature.

5

THE ROLE OF MACROPHAGES IN THE METABOLISM AND DISPOSITION OF IRON, LIPIDS, STEROIDS AND PROTEINS IN NORMAL AND PATHOLOGICAL CONDITIONS

IRON METABOLISM

The total body content of iron in the normal adult human is about 4 g (range 2–6 g). Less than 0.1 per cent of the total body iron is composed of free ferrous ions and protein-bound iron in transport in the plasma. The remainder is either bound to a porphyrin ring as part of blood or muscle haemoglobin or as one of the haem enzymes (cytochromes, catalase and peroxidase), or is iron in storage form. The storage forms of iron, ferritin and haemosiderin, constitute about 30 per cent of the total iron content of the body, which tends to remain fixed within fairly narrow limits, otherwise a state of iron deficiency or haemosiderosis occurs. Thus any loss of iron from the body must be replenished by the absorption of iron from the diet.

Iron enters the plasma, not only by absorption from the gut but also from the breakdown of haemoglobin and from the release of storage iron. About 27 mg of iron enter and leave the plasma each day. Of this amount, about 21 mg are derived from the breakdown of effete red cells, and the remainder is largely derived from released storage iron, with a small percentage only contributed by dietary iron. In the plasma, iron is transported by a specific plasma protein called transferrin (siderophilin). Normally, about one-third of the transferrin is saturated with iron. In addition to the transferrin circulating in the plasma, transferrin is also sequestered within certain cells, and macrophages and some reticulum cells have the particular capacity to retain a large proportion of cellular transferrin. Macrophages can also synthesize transferrin. Iron is stored in two forms, ferritin and haemosiderin. The liver, spleen, bone marrow, kidneys and, in some animals, the intestinal mucosa, are all rich in ferritin. Haemosiderin is largely confined to macrophages in the spleen, bone marrow and other organs, and, unlike ferritin, under normal circumstances is present in very small amounts in the parenchymal cells of the various organs. It may be concluded that intermediary iron metabolism in the normal individual is achieved largely by the rapid interchange between the two mechanisms of proteinic transport – transferrin in the plasma and macrophages, and ferritin in the parenchymal cells of various organs. Under normal circumstances, the iron in the plasma has a half-life between one and two hours, and labelling studies demonstrate that most of it is incorporated into the haemoglobin of developing red cells within a few days.

ERYTHROPOIESIS

During erythropoiesis, it seems likely that the iron of plasma transferrin can be directly transferred to form the ferritin in developing red cells in the bone marrow, probably during the pro-erythroblast stage, and iron is transported to the mitochondria where haemoglobin is synthesized. The excess iron, in the form of ferritin, is returned directly to macrophages in the bone marrow without re-entering a plasma phase. It seems likely that the greater part of the iron needed for the formation of haemoglobin is derived from plasma transferrin in this way. However, an alternative mechanism may exist whereby iron stored in macrophages and reticulum cells in the bone marrow is transferred directly to erythroblasts during the process of 'rhopheocytosis'. Bessis and Breton-Gorius (1962) have drawn attention to the presence of islands of erythroblasts in the bone marrow which are always closely grouped around large cells of the macrophage-reticulum type. The central cell of each island, containing stored ferritin, appears to supply the erythroblasts with ferritin directly by a process in which ferritin from the cytoplasm of the central cell adheres to the plasma membrane of the erythroblast. The ferritin becomes transported within the cytoplasm of the erythroblast by invagination of the cell membrane to form small vesicles containing ferritin by a process similar to pinocytosis. Later the membrane surrounding the vacuole disappears, the fluid within the vacuole is absorbed, and the ferritin passes to mitochondria where it is available for haemoglobin synthesis. The extent to which this process (rhopheocytosis) contributes in the overall transfer of iron to erythroblasts is uncertain, and it would appear that the greater part of the iron needed for the formation of haemoglobin is derived from plasma transferrin. It has been suggested by some authors that rhopheocytosis is the means whereby excess iron is transferred from erythroblasts to the central storage cell (i.e. the ferritin is passing in the reverse direction), but Bessis (1963), deducing a dynamic process largely on the basis of the static conditions of electron microscopy, appears to entertain little doubt that the direction of iron transfer is from the central cells to the erythroblasts.

Bessis (1963) has also described how, during the formation of the reticulocyte, the nucleus of the erythroblast is expelled together with a surrounding rim of haemoglobinized cytoplasm in 90 per cent of cases. The whole of the Golgi region of the erythroblast may be contained within this ring of expelled cytoplasm. The expelled nucleus, fragmented in some cases, together with its ring of cytoplasm, is then ingested by macrophages in the bone marrow. Bessis calculated that 5 to 10 per cent of reticulocyte haemoglobin is normally lost from the red cell during the phagocytosis of erythroblast nuclei. The haemoglobin ingested during this process of nucleophagocytosis is broken down by the macrophage to release iron and bilirubin. The iron is re-utilized by being incorporated anew into ferritin and haemosiderin, and the bilirubin is released into the plasma and is subsequently taken up by the liver. The amount of haemoglobin which is taken up by macrophages during reticulocyte formation is small in comparison with the amount ingested and catabolized by macrophages during the removal of effete erythrocytes from the circulation during the normal process of erythrophagocytosis (erythroclasia); and this latter process provides about 75 per cent of the iron which enters and leaves the plasma each day.

ERYTHROPHAGOCYTOSIS (ERYTHROCLASIA)

It has long been known that macrophages remove senescent, damaged or dead erythrocytes from the circulation. A very large literature has accumulated on various aspects of this mechanism, but precise knowledge of some aspects of the phenomen is still lacking at this time. While there is no doubt that macrophages are the cells primarily responsible for the removal of effete erythrocytes from the circulation, the means whereby these cells discriminate between healthy erythrocytes and the effete cells is largely speculative. It has been shown that guinea-pig macrophages, which show no tendency to interact *in vitro* with fresh autologous red cells, have a marked affinity for 'aged' autologous red cells; and the adherence of aged red cells to the macrophages does not appear to be dependent on the prior adsorption on to the red cell surface of serum factors (Vaughan and Boyden, 1964). Further studies by Vaughan (1965b) suggest that the interaction between macrophages and aged autologous red cells may be due to the presence on the macrophage of a natural cytophilic antibody. He observed that aged autologous red cells did not adhere to trypsin-treated macrophages unless the latter had been incubated with fresh autologous serum. The serum factor responsible for the adherence phenomenon could be removed by adsorption with aged autologous red cells but not by fresh autologous red cells. In contrast, Nelson (1969) was unable to find evidence of a cytophilic antibody which is both removed by trypsin and capable of re-attachment to trypsinized macrophages in experiments with guinea-pig antisera to sheep erythrocytes.

It is of interest to note that Stuart and Cumming (1967) found that mouse macrophages grown in human serum were able to discriminate between fresh and effete human red cells. The failure of mouse macrophages to phagocytose the fresh human red cells was taken to suggest that the maintenance of macrophages in human serum deprived them of natural opsonins for the red cells of this species. Stuart (1970) also reports that human macrophages show no capacity to interact with homologous red cells in the presence of human serum lacking normal blood group antibodies (AB serum), but immune anti-A or anti-B sera both produce marked erythrophagocytosis. The preceding observations suggest that a naturally-occurring serum factor behaves in the manner of a cytophilic auto-antibody which is specific for some determinant present (or revealed) only on aged erythrocytes (Nelson 1969).

The phagocytosis of intravenously injected homologous and heterologous erythrocytes has been extensively studied as a means of examining the action of various factors (antibody, complement, haemolysins, etc.), but the relevance of the findings to erythrophagocytosis under normal conditions is largely speculative.

A major question concerning erythroclasia is the role of the spleen in this process. Histological and chemical examination shows that the liver, the spleen and the bone marrow are all particularly rich in iron content. Various attempts to show a decrease in red cell concentration in the splenic vein relative to the splenic artery have not been helpful. Moreover, although splenectomy may increase red cell survival in certain pathological conditions of excessive haematoclasia, the removal of this organ does not do so in normal individuals. Red cell numbers and red cell survival are the same in splenectomized subjects as in normal individuals. However, it is of interest to note that there is usually a lasting increase in the number of circulating leucocytes and platelets after splenectomy. The findings suggest that a special role in normal erythrophagocytosis cannot be attributed to the spleen, and some investigators hold the view that the bone marrow plays the major role in normal erythrophagocytosis in the human (Ehrenstein and Lockner, 1958; Bessis, 1963).

It has been reported that, immediately prior to its phagocytosis by macrophages, the red cell generally breaks up into two portions, and after ingestion divides into further small fragments in a very short space of time (Bessis, 1963). The breaking up of effete red cells prior to phagocytosis may be the result of the action of non-specific and specific lysins. Bergenhem and Fåhraeus (1936) expressed the opinion that lysolecithin acts on red cells as a haemolysing factor. The lysolecithin may be formed under the action of serum lysolecithinase, especially in regions of blood stasis as in venous sinusoids in the spleen and bone marrow. In this connection, it is of interest to note that macrophages are capable of synthesizing phospholipids, including lecithins. Under such circumstances, it is postulated that the cell membrane of the aged red cell is more susceptible to lytic factors than the young cells. Hyaluronidase, lysozyme and β-glucuronidase also possess haemolytic activity, and these enzymes are present in macrophage lysosomes. A deterioration in the strength and elasticity of the cell membrane of senescent red cells has been put forward as a reason why the aged cells could undergo disruption as a result of the mechanical trauma of normal blood flow, and particularly in the process of squeezing between the lining cells of the splenic sinusoids in order to enter the cord spaces. A similar mechanism of mechanical 'weakness' has been postulated by some authors as being the reason for the premature phagocytosis of abnormally-shaped erythrocytes in spherocytosis.

In contrast to the above findings, there is other evidence that red cells may not be broken down prior to their ingestion by splenic macrophages. Edwards and Simon (1970) studied the destruction of red cells in the normal rat spleen with the aid of the electron microscope, and could find no morphological evidence of intravascular lysis or extracellular fragmentation of red cells, and no evidence of extravascular degeneration of red cells was found. They found that the macrophages in the red pulp, the marginal zone and peripheral white pulp were involved in erythrophagocytosis, but the cells lining the venous sinuses were not. After macrophages had endocytosed the intact erythrocytes, the membranes of the phagosomes invaginated into the endocytosed erythrocytes to form a system of interconnecting tunnels which became occupied by the cytoplasm (and some organelles) of the macrophage. During this process, much ferritin accumulated in the walls of the tunnels and eventually passed into the surrounding cytoplasm. Extensive 'tunnelization' eventually caused the breaking up of the endocytosed erythrocytes into small fragments.

In conclusion, it may be stated that although the precise mechanism of erythrophagocytosis is unknown, the macrophages of the liver, spleen and bone marrow are the principal cells involved under normal conditions. Macrophages are also capable of producing factors which may have a lytic action on the senescent red cell, and may possess the means of actively discriminating between senescent red cells to be phagocytosed and healthy cells to be allowed to remain in the circulation.

PATHOLOGICAL ERYTHROPHAGOCYTOSIS

In certain pathological states there is excessive erythrophagocytosis. Under such circumstances, the role of the macrophages may be a relatively active or passive one. The macrophages may play a selectively passive role during the excessive phagocytosis of red blood cells which are 'abnormal' due to reasons of corpuscular or extra-corpuscular origin. Conversely, in conditions where there is hyperactivity of the macrophage system, normal healthy red cells may be removed from the circulation during excessive erythrophagocytosis.

Excessive erythrophagocytosis may be observed in the conditions characterized by inherited anomalies in the red cells, such as in sickle-cell anaemia, Cooley's anaemia and hereditary haemolytic jaundice. Red cell anomalies may also be acquired, as during pernicious anaemia and in auto-immune haemolytic anaemias. In all these conditions, the macro-phage system apparently functions normally but rapidly removes the abnormal cells from the circulation so that the overall red cell survival time is reduced. There is some evidence that the spleen may play a special role in certain of these conditions. Removal of the spleen generally has no effect on the haematological course of the disease in sickle-cell and Cooley's anaemia, but may be beneficial in hereditary haemolytic jaundice. Thus it appears that the degree of erythrophagocytosis in the spleen de-pends upon the nature of the red cell abnormality. It is of interest to note that, if a patient with hereditary haemolytic jaundice is transfused with normal red cells, the survival of the transfused red cells is normal; however, if a normal subject is transfused with red cells from a patient with hereditary haemolytic jaundice, the transfused abnormal cells are rapidly eliminated in the normal spleen (Dacie and Mollison, 1943).

The macrophage system may also function normally during the excessive erythrophagocytosis which characterizes various haemolytic anaemias due to various 'extrinsic' extracorpuscular factors acting on red cells. While non-specific factors may cause some changes in the red cells during uraemia and during general toxic states, and thereby hasten their removal from the circulation, the most important group of conditions in this category are the idiopathic auto-immune haemolytic anaemias. The anti-bodies responsible for promoting the destruction of normal red cells in auto-immune haemolytic anaemias may be of several types. They may be 'warm' antibodies (usually 7S γ-globulins – IgG) optimally active at 37 °C and not usually complement-fixing, or 'cold' antibodies (19S γ-globulins – IgM) optimally active at temperatures below 37 °C and usually complement-fixing. In both cases, the antibodies may be 'incom-plete' and can then only be detected by the (Coombs) antiglobulin test. In most cases, the warm antibodies are of unknown specificity, but in about one-third of cases warm antibodies may exhibit specific reactivity against human blood group antigens. Cold auto-antibodies usually show reactivity against the widely distributed I antigen. In addition to the ideo-pathic groups of auto-immune haemolytic anaemias, haemolytic anaemias with demonstrable antibodies may be present in syphilis, malignant lymphomas and systemic lupus erythematosus.

It would appear that, in the auto-immune haemolytic anaemias, the antibodies do not promote intravascular haemolysis to any marked degree but act as opsonins to promote the phagocytosis of the intact red cells. The therapeutic effects of splenectomy in these anaemias is variable, and this may be due to the fact that erythrocytes with small amounts of antibody on their surfaces are taken up predominantly by the macrophages of the spleen, but those cells with larger amounts of antibody are predominantly taken up by the macrophages of the liver. Moreover, in some cases red cell survival may be little below normal when antibody levels are high, and excessively accelerated erythroclasia may be present when antibody levels are low. Demonstrable antibody is not always present continuously, and re-activation of haemolysis is rarely heralded by a demonstrable rise in antibody levels. These observations suggest that other immunological or non-immunological factors, possibly involving the activity of macrophages responsible for erythroclasia, act to regulate the rate of red cell destruction in these diseases. In this connection, it is known that, in the auto-immune haemolytic anaemias, the red cells become agglutinated to form clumps within the splenic sinusoids, and many red cells are also transformed from the normal disc-like shape towards a more spherical form; these factors would seem to render the red cells more liable to mechanical damage and phagocytosis.

Under certain conditions, notably in some drug-induced allergic reactions, haemolytic anaemias having an immunological basis may also be observed. These haemolytic anaemias are not auto-immune and are only operative in the presence of the drug. The drug appears to act as a hapten (protein-free substance which can interact with specific receptor groups on an antibody molecule but cannot itself elicit antibody formation) which unites with the patient's red cells (the carrier) in the presence of the patient's serum (Ackroyd and Rook, 1963). The antibodies formed are hapten-specific (i.e. drug specific) but their interaction with the hapten antigen (the drug) leads to damage to the carrier red cells which are then removed from the circulation by macrophages, and particularly by macrophages within the spleen.

There remains a group of haemolytic anaemias in which auto-antibodies cannot be demonstrated. Some of these conditions have been attributed to 'hypersplenism' implying a state of splenomegaly associated with increased erythroclasia by the macrophages of that organ. Haemolytic anaemia associated with splenomegaly may be seen in a heterogeneous group of conditions including primary and secondary malignant neoplasms involving the spleen, myeloproliferative diseases, lipid storage diseases, Felty's syndrome, Banti's syndrome, brucellosis, tuberculosis of

the spleen, typhoid, and other conditions. Attempts to demonstrate antibodies in the serum or on the red cells have failed, but it is possible that
there exist cytophilic auto-antibodies bound to macrophages and not
present in serum or on red cells (Nelson, 1969). It is known that, in cases
of splenomegaly, large volumes of blood may be sequestered within the
enlarged organ; this pooling effect may allow haemolytic agents to be
effective on trapped normal or minimally damaged red cells and lead to
their ingestion by macrophages. However, it appears to be quite clear that
although the enlarged spleen plays some part in the haematological
abnormalities in those conditions accompanied by hypersplenism, the
underlying mechanism is probably more complex and other factors and
other components of the macrophage system may be involved. For
example, studies on red cell survival in experimental tuberculosis in
guinea-pigs indicate that the macrophages of both spleen and liver may be
involved in taking up healthy and aged red cells from the circulation, and
that this may be partly due to an increase in the phagocytic activity of
macrophages throughout the body in this condition (Miescher, 1957).
Excessive erythroclasia resulting in severe anaemia has been observed
during the stimulation of macrophage activity in mice using zymosan
(Gorstein and Benacerraf, 1960) and in rats using *p*-dimethyl-amino-
benzene (Lozzio, Machado and Lew, 1966).

HAEMOSIDEROSIS AND HAEMOCHROMATOSIS

In haemosiderosis increased quantities of iron are found in the macrophages and, to a lesser extent, the parenchymal cells in various organs.
The liver, spleen, kidneys and bone marrow are the principal organs
involved, and the affected tissues are a rusty brown in colour. The liver,
although heavily impregnated with iron, usually retains its normal lobular
architecture, unlike in haemochromatosis when the liver often becomes
cirrhotic. In most cases of haemosiderosis the excess iron in the body is
due to the excessive breakdown of haemoglobin or to excessive iron
administration. Thus an excess of haemosiderin may be found within
macrophages locally as a result of trauma producing localized haemorrhage and bruising, and generally as a result of prolonged and excessive
therapeutic iron administration by the parenteral routes, repeated blood
transfusions involving several hundred pints over a long period, and
excessive blood destruction in haemolytic anaemias from any cause.
Excess haemosiderin may also be present in alveolar macrophages in the
common condition of congestive heart failure (Fig. 20) and in the very
rare condition of ideopathic pulmonary haemosiderosis. Excessive quantities of haemosiderin are also deposited in the tissues in the condition of
haemochromatosis.

Haemochromatosis is a rare inborn error of iron metabolism which, like haemosiderosis, is associated with the excessive accumulation of iron in the tissues. There is an increased absorption of iron from the alimentary tract, the mechanism of which is unknown. The distribution of iron pigment differs from that in haemosiderosis in that, in haemochromatosis, the hepatic parenchymal cells are more heavily laden than the Kupffer cells, and there is also a marked tendency to fibrosis in the affected tissues. Thus haemochromatosis leads to liver failure due to cirrhosis (fibrosis), to diabetes mellitus due to destruction of the pancreas, and to excessive pigmentation of the skin.

LIPID METABOLISM

There is much experimental and pathological evidence that macrophages are involved in the metabolism of lipids (including cholesterol) of endogenous and exogenous origin. Much of the experimental work has been carried out in relation to the pathogenesis of atherosclerosis and, to a lesser extent, hypercholesterolaemia and lipid storage disorders.

UPTAKE OF LIPIDS BY MACROPHAGES

Nearly all dietary lipid enters the blood stream from the chyle of the thoracic duct in the form of small particles known as 'chylomicrons'. The chylomicrons are essentially composed of a central core of triglyceride, associated phospholipid and with free and ester cholesterol, the former serving as a surfactant material. Monoglycerides, diglycerides and free fatty acids may also be present. About 80–90 per cent of chylomicron lipid is composed of triglyceride which has been synthesized in the small intestinal mucosa from dietary fatty acids. The protein which is adsorbed on the surface of the particle endows it with a negative charge.

There is general agreement that chylomicrons are removed from the blood stream as an intact unit (DiLuzio and Riggi, 1967), and intravascular hydrolysis accounts for the disappearance of less than 5 per cent of this lipid. Studies on the removal of chylomicrons of endogenous and exogenous origin from the blood stream have suggested that the dynamics of this process may resemble that of the phagocytosis of other types of particles by macrophages following intravenous injection. In this connection, it has been reported that intravenously injected labelled chylomicrons disappear from the blood at an exponential rate (Havel and Fredrickson, 1956; French and Morris, 1957). French and Morris (1957) found that the disappearance curve for chylomicrons closely resembled that of inorganic colloids in that, probably due to variations in particle size, the curve revealed the presence of a fast and slow component, the larger chylomicrons disappearing faster than the smaller ones. Byers (1960) has pointed out that this latter observation is easy to account for if the macrophages are responsible for chylomicron clearance but difficult to explain on another basis. For instance, intravascular hydrolysis or another enzymatic process should act more rapidly on smaller particles. Similarly, if chylomicrons gain access to hepatic cells by passing through the walls of the liver sinusoids via the gaps (pores) between Kupffer cells, the smaller particles should pass through the pores more easily than the larger ones since they are less likely to collide with the boundaries of the gap. While this would appear to provide a logical explanation for the chylomicron clearance curves, further experimental evidence indicates that it is the hepatic parenchymal cells which are primarily concerned in the clearance of chylomicrons from the blood, although the Kupffer cells may still play an important role in other aspects of lipid metabolism.

The administration of an artificially prepared triglyceride emulsion intravenously to rats is followed by its removal almost entirely by the liver and it can be detected within both Kupffer cells and parenchymal cells; however, blockade of the Kupffer cells induced by inorganic particulates does not significantly alter the rate of removal of emulsified fat from the blood although its appearance within Kupffer cells is prevented, and the liver parenchymal cells still contain fat under these circumstances (Waddell, Geyer, Clarke and Stare, 1954). Moreover, histological examination of the liver after perfusion with ^{14}C-tripalmitin-labelled chylomicrons has revealed that most of the fat is located in the parenchymal cells with some fat present within Kupffer cells (Morris and French, 1958). Thus it would seem that the mechanism of direct particulate uptake by the liver parenchymal cell, in which the Kupffer cell is largely bypassed, is by way of the sinusoidal pores. Pore sizes of up to 2μ in diameter are present, and this would allow the direct passage of most chylomicrons into the space of Disse. Studies on the size of chylomicrons in rats fed on either butter or corn oil (Jones, Thomas and Scott, 1962) revealed that the chylomicrons in the butter-fed rats ranged from 50 mμ to 20μ. The chylomicrons in excess of 15μ were classified as 'giant' chylomicrons. No 'giant' chylomicrons were seen in corn-oil-fed rats. The observations of French and Morris (1957) that large chylomicrons are removed more rapidly than smaller ones, suggests that the large chylomicrons, unable to pass through sinusoidal pores, may be phagocytosed by Kupffer cells.

From all the available evidence, some of which has been cited above, it would appear that, under normal circumstances, macrophages have, at most, a minor role in the removal of chylomicrons from the circulation. However, the extensive studies of DiLuzio and his colleagues have shown that, in contrast to chylomicrons, intravenously injected synthetic lipid emulsions are removed from the circulation exclusively by macrophages, particularly the Kupffer cells (Ashworth, DiLuzio and Riggi, 1963), and are suitable for evaluating the functional activity of macrophages in experimental animals and clinical subjects. Thus it is evident that the results obtained using synthetic lipid emulsions cannot be applied in the interpretation of normal chylomicron metabolism.

Although macrophages may play a minor role in chylomicron removal from the circulation, there is evidence that some dietary cholesterol may be removed from the circulation by Kupffer cells. Friedman, Byers and Rosenman (1954) demonstrated that cholesterol was present in Kupffer cells 6 hours after its oral administration, and was present in larger amounts in Kupffer cells at 24 hours. 'Blockade' of macrophages produced with foreign colloids (carbon, chromium phosphate, saccharated iron) resulted in reduced cholesterol deposition in both Kupffer and liver parenchymal cells, and hypercholesterolaemia persisted for a longer period than in non-blockaded control animals. Subsequent studies with ^{14}C-labelled cholesterol were interpreted as indicating that dietary cholesterol is removed from the blood by Kupffer cells and is then transferred to the liver parenchymal cells for metabolic purposes (Byers, Mist-St George and Friedman, 1957). However, the subsequent studies of DiLuzio appear to be in direct contradiction to the above conclusions. DiLuzio (1959) examined the chemical composition of separated liver parenchymal cells and Kupffer cells, and reported the presence of higher levels of free and esterified cholesterol in the Kupffer cells. DiLuzio (1960) also reported that in rats, maintained on high-cholesterol diets, hyperplasia and hyperfunction of macrophages induced by zymosan was accompanied by a profound lowering of liver and plasma cholesterol. Studies on the cellular distribution of administered cholesterol indicated that the major site of removal was the hepatic parenchymal cell, but a subsequent elevation in Kupffer cell cholesterol content appeared to indicate a metabolic or excretory function of these cells in cholesterol metabolism.

DiLuzio and Riggi (1967) have recently shown that the intravascular removal rate of cholesterol-containing and triglyceride-containing chylomicrons is not enhanced by stimulating the phagocytic activity of macrophages, confirming that the Kupffer cells do not play a significant part in the uptake of lipids in this form. These latter studies also established that the great majority of ^{131}I-labelled triglyceride or ^{14}C-cholesterol-containing chylomicrons are taken up initially by hepatic parenchymal cells, and the uptake in spleen and lungs is relatively minor, being in the order of 1–3 per cent of the administered dose at 10 minutes after injection. Byers *et al.* (1957) also found no significant localization of cholesterol in spleen and lung after feeding cholesterol to rats. These findings are in direct contrast to the marked uptake of lipid by lung macrophages which may be observed after the oral or intravenous administration of some synthetic lipid emulsions, and which is not observed when physiological emulsions are employed (DiLuzio and Riggi, 1967). However, it is interesting to note that Bernick and Patek (1961) observed the accumulation of neutral lipid, but not cholesterol, in alveolar macrophages in long-term studies on rats receiving a high-fat, high-cholesterol diet.

METABOLISM OF LIPIDS BY MACROPHAGES

The extensive studies of Day and his colleagues have done much to elucidate the role of macrophages in various aspects of lipid metabolism (reviewed by Day, 1964) and a brief outline of some of their findings follows here. Rabbit macrophages contain cholesterol esterase activity, both synthetic and hydrolytic, and macrophages which have taken up cholesterol or cholesterol oleate are capable of the synthesis and hydrolysis of cholesterol ester. Macrophages also contain other esterases and lipase and are capable of partly oxidizing chylomicron fat, triglyceride and fatty acid taken up by the cells *in vitro*. Following the uptake of ^{14}C-labelled fatty acids *in vitro* by macrophages, most of the fatty acid is incorporated into triglyceride and phospholipid, with some incorporation into cholesterol ester and mono- and diglycerides. The uptake of ^{14}C-cholesterol suspension by macrophages *in vitro* is followed by the return of cholesterol into the medium in a soluble form (presumably a lipoprotein complex) which is more readily esterified and more readily incorporated into serum lipoprotein in the medium. The uptake of cholesterol is accompanied by an increase in phopholipid synthesis; however, phospholipid decreases the esterification of free cholesterol by macrophages, although it also promotes the hydrolysis of lipoprotein cholesterol.

It is of interest to note in relation to atherosclerosis that esterification with predominantly saturated fatty acids occurs when cholesterol in suspension is ingested by macrophages. When corn oil or coconut oil are ingested together with cholesterol by the macrophages, the fatty acid pattern can be influenced towards the polyunsaturated or saturated side respectively.

Gas phase chromatographic studies have shown that macrophages *in vitro* can also incorporate ^{14}C-labelled acetate into palmitic, oleic and myristic acid, smaller amounts being incorporated into other long-chain fatty acids.

The above findings of Day and his colleagues, together with other available evidence, show that macrophages from various sites contain a variety of enzymes associated with lipid metabolism and are able to synthesize and hydrolyse cholesterol ester, to oxidize fatty acid and triglyceride, to synthesize phospholipid, and to synthesize fatty acids from acetate.

Marked changes in lipid metabolism in macrophages occur during phagocytosis. The incorporation of acetate and glucose carbon into lipids is significantly increased and the incorporation of ^{32}P into complex phosphatides is also increased (Karnovsky *et al.* 1966; Oren *et al.* 1963). It is tempting to suggest that these changes reflect the need to synthesize phospholipid-containing cell membrane components to replace that part of the cell membrane interiorized during the phagocytic act. However, these observations must be extrapolated with caution since the findings were shown to accompany phagocytosis by peritoneal macrophages, but not by alveolar macrophages.

MACROPHAGES AND ATHEROSCLEROSIS

Atherosclerosis is the commonest, the most important and the most controversial of arterial diseases. It is so universal that it may be regarded with some justification as a normal concomitant of ageing, and it is the principal pathogenic factor in two of the commonest causes of death in the middle-aged and elderly, namely myocardial infarction and cerebral infarction. The disease principally affects, in an irregular manner, large and medium-size arteries.

Three stages may be considered in the development of the arterial lesions. They are (1) fatty streaking, (2) plaque formation, and (3) fully-developed atherosclerosis. These stages may all be present in adjacent portions of the same artery. Microscopically, the early lesion of fatty streaking is composed of a subendothelial collection of large 'foamy' macrophages containing abundant lipid material (Fig. 34), much of which can be demonstrated to be cholesterol. Microscopic examination of discrete atheromatous plaques reveals a sizeable collection of lipid lying within and thickening the intima. The lipid is still partly contained within macrophages, but in the central part of the lesion the cells have often disintegrated to liberate a structureless fatty mass with crystals of cholesterol present. The overlying intima is thickened, while the underlying internal elastic lamina may be fragmented and the media reduced in thickness. Finally, there is widespread proliferation of fibrous tissue around and between plaques, so that calcification, ulceration, mural thrombosis and occlusive thrombosis may ensue in the affected area, and the wall of the affected vessel may be weakened by the disease process.

There are two classical theories regarding the aetiology of atherosclerosis: the 'infiltration' theory ascribes the greater significance to the accumulation of lipids in the intima by infiltration from the blood, while the 'thrombogenic' theory maintains that the formation and organization of mural deposits and thrombi constitute the major process. Other theories also exist, and the problem of the underlying pathogenesis of this

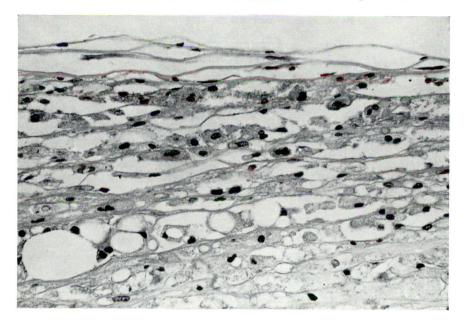

Fig. 34. Lipid-containing macrophages ('lipophages') in intima of human aorta in an early atheromatous plaque. Haematoxylin–eosin. × 1,000.

condition has attracted a vast literature. There exist many books and review articles dealing with various features of the atherosclerotic lesion in man and experimental animals, and Day (1964) has specifically reviewed the possible role of macrophages in atherosclerosis.

Whatever the aetiology of atheroma may be, it is generally agreed that macrophages containing much lipid within their cytoplasm ('lipophages') are a prominent feature of atherosclerotic lesions in man and experimental animals, and they may be readily distinguished from the lipid-containing smooth muscle cells which also may be present, using the electron microscope (French, 1966). It has been reported that in the rabbit macrophages can proliferate within the intima (Poole and Florey, 1958), and tritiated thymidine studies have confirmed that these cells are actively engaged in DNA synthesis (Spraragen, Bond and Dahl, 1962). The possible origin of the macrophages which appear in atherosclerotic lesions has been reviewed by French (1966). Occasional macrophages occur in the intima of normal large arteries, and these could proliferate in atheroma by cell division, but the greater concentration of macrophages towards the edges and surface of lesions is suggestive of an influx from the blood. In this connection, Poole and Florey (1958) showed that in rabbits fed with cholesterol, lipid-containing macrophages adhered to the endothelium of the aorta and migrated through it into the intima. Simon, Still and O'Neal (1961) have also shown by electron microscopy the accumulation of lipid-containing macrophages in the aortic intima in rats fed with a high-fat diet: similar cells were also demonstrated in the circulating blood and actively penetrating the aortic endothelium. Thus, under these experimental conditions of a very high lipid intake, there is evidence that the macrophages which accumulate in the intima could have acquired at least some of their lipid beforehand. However, it can also be shown that not all the macrophages within early atherosclerotic lesions contain demonstrable lipid.

Adams and Tuquan (1961) have reported that histochemical studies on early human atheromatous lesions reveal that most of the lipid in the early plaque is phospholipid, possibly resulting from the degeneration of elastic tissue at the intimo-medial junction. This phospholipid first appeared extracellularly and not within macrophages, which were shown to contain cholesterol but little phospholipid. These authors suggested that this lipoidal degeneration of elastic tissue, with pooling of phospholipid, precedes the deposition of hydrophobic lipid from the blood during the formation of the atheromatous plaque. However, Dunnigan (1964) also studied human atherosclerotic plaques by histochemical methods, and demonstrated that a significant amount of phospholipid is present within a variable proportion of macrophages in the atherosclerotic plaque. The location of this phospholipid on the intimal side of the plaque and within macrophages suggests that it is distinct from the phospholipid demonstrated at the intimo-medial junction by Adams and Tuquan (1961). Day (1964) has also shown that phospholipid is present entirely within intimal macrophages in aortic lesions in cholesterol-fed rabbits. Biochemical studies have shown that experimentally deposited cholesterol is removed, presumably by local phospholipid synthesis; and phospholipids are synthesized *in situ* in the lesions in experimental animals and man (Zilversmit *et al.* 1961).

Macrophages are capable of synthesizing phospholipid from fatty acids or acetate (Day, 1964), and taking into account the close proximity of phospholipid-containing macrophages to hydrophobic lipids in atheromatous plaques, this suggests that macrophages may be involved in the transformation of deposited hydrophobic lipids to phospholipids. Moreover, if the intimal macrophages possess the functional properties in lipid metabolism cited previously, it seems that their activities could modify the composition of lipids derived from plasma lipoproteins or other sources and perhaps explain some of the complex changes which have been observed in the lipid composition of the atherosclerotic lesion (French, 1966), and there is abundant evidence that serum lipoproteins (including cholesterol) may infiltrate from the lumen into the aortic intima.

The experimental production of arterial lesions in animals given intravenous doses of colloidal suspensions or macromolecular substances other than cholesterol has been demonstrated. Patek, Bernick and Frankel (1961) have shown that rats given a single injection of colloidal carbon did not show coronary artery lesions until five months after the injection, when these vessels exhibited some fragmentation of the internal elastic lamina and intimal thickening. Stainable lipid was not demonstrable in these arterial lesions. Rats given a standard diet with 1 per cent cholesterol and 5 per cent fat added to it, demonstrated no vascular lesions even after 7 months on this diet. However, rats placed on the cholesterol–fat diet three days after carbon injection exhibited atherosclerotic lesions in the coronary arteries and aorta as early as 2 months afterwards. Moreover, rats maintained on the cholesterol–fat diet for 5 months and then given a single injection of carbon exhibited coronary artery atheroma as early as 21 days after the carbon injection. Further studies by Patek, Bernick and de Mignard (1967) suggested the existence of a cyclic process of release, re-circulation and re-phagocytosis of carbon which proceeds indefinitely following its initial uptake by macrophages, and that the repeated uptake of particulate matter by macrophages causes them to release a substance capable of producing arterial lesions, with the secondary accumulation of lipid within the macrophages located in the damaged vessel wall. Other studies have shown that 'blockading' doses of various colloids administered intravenously, or the administration of substances known to depress the phagocytic activity of macrophages, such as corticosteroids or Tween 80, enhance hyperlipaemia and hypercholesterolaemia and accelerate the production of atherosclerosis in animals fed on normal or high-fat diets (Antonini, Weber and Zampi, 1960; Rosenman, Breall and Friedman, 1960). While these findings do not necessarily indicate a role for the macrophage in the normal structural stability of the arterial wall, there is little doubt that disturbances in lipid metabolism can follow the suppression of macrophage function under experimental conditions.

Phagocytosis of cells or cell debris by macrophages at sites of thrombosis or intimal haemorrhage has been put forward as an explanation for some of the features of atherosclerotic lesions. Haemosiderin is frequently seen in advanced atherosclerotic lesions in man, and electron microscopic studies of organizing experimental thrombi have demonstrated the phagocytosis of red cells and red cell fragments by macrophages, and the subsequent appearance of ferritin within the cytoplasm of these cells (Wiener and Spiro, 1962).

There is now much evidence that organization of mural thrombi in arteries plays an important part in the pathogenesis of atherosclerosis. Poole (1966) showed by electron microscopy that macrophages phagocytose platelets in the mural thrombus which forms on the surface of fabric prostheses in the baboon aorta, and similar findings have been reported by others. Hand and Chandler (1962) studied the fate of artificial thrombi when injected intravenously in rabbits and when incubated *in vitro*, and obtained light microscopical evidence that platelets were engulfed by macrophages: subsequently those macrophages which had ingested platelets showed stainable lipid within their cytoplasm and became morphologically indistinguishable from the 'lipophages' (macrophages laden with fat droplets) which can be seen in many atherosclerotic lesions. However, the ingestion of platelets would seem unlikely to account for more than a minor proportion of the total lipid content of such lesions under normal circumstances. The role of macrophages in the disposal of fibrin is uncertain. Lee and McCluskey (1962) demonstrated that the microprecipitates of fibrin, formed following the intravenous injection of thrombin, are removed from the circulation by the Kupffer cells of the liver. Barnhart and Cress (1967) have also shown that macrophages remove circulating aggregates of fibrin and fibrinogen from the blood, and the Kupffer cells are especially effective in the clearance of circulating soluble derivatives of fibrinogen and fibrin which can result from proteolysis.

XANTHOMATOSIS

The presence of clinically recognizable deposits of cholesterol and other lipids in the skin, tendons, bone, periosteum and joints occurs in many diseases. Most of the lipid in these deposits is contained in large foamy or granular macrophages (Fig. 35), and the individual aggregates are known as 'xanthomata'. Plaques of xanthomata around the eyes ('xanthelasma') are commonly encountered in elderly persons, often with no demonstrable abnormality in blood lipids. Moreover, xanthomata are also commonly encountered in cases where hyperlipidaemia is secondary to some well-defined disease, as in cases of diabetes mellitus, myxoedema, alcoholism, etc. There remain those cases where xanthomatosis is associated with a primary disorder of lipid metabolism, often clearly familial, and not associated with any recognizable underlying disease.

The hyperlipoproteinaemias associated with xanthomatosis have been classified into five types on the basis of electrophoretic lipoprotein analysis (Fredrickson, Levy and Lees, 1967). Xanthomata may accompany all five types of hyperlipoproteinaemia, and vary in size, number and distribution (Fig. 36). The level of lipids in the serum may bear little relation to the age of appearance of xanthomatosis and the number of lesions present. Moreover, eruptive xanthomata of the skin may appear in crops and disappear again over the course of a few days in both primary and secondary

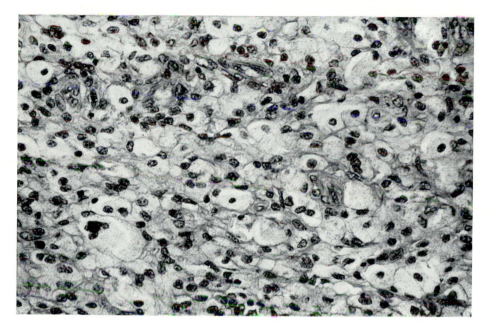

Fig. 35. Xanthoma cells showing foamy cytoplasm and multinucleate forms. Haematoxylin–eosin. × 1,000.

hyperlipidaemias. The premature development of atherosclerosis and its complications occurs during types II and III hyperlipoproteinaemia, but its association with types IV and V is obscured by the common complications of diabetes and obesity.

THE LIPOIDOSES

The lipoidoses are an uncommon group of disorders arising from congenital abnormalities of lipid metabolism, usually inherited as autosomal recessive characteristics, and giving rise to focal or diffuse tissue lesions composed of granular or foamy lipid-rich macrophages.

Gaucher's disease is the commonest of the lipoidoses. There is an infantile form of the disease exhibiting signs of involvement of the nervous system about six months after birth, and running an acute and fatal course. The more frequent adult form may be discovered at almost any age, and about half of the total number of cases do not present clinically until after the age of 30 or later. The principal clinical features are gross enlargement of the spleen to about ten times normal size, hepatomegaly, lymph node enlargement, pingueculae in the eyes, and abnormal melanin pigmentation of the skin. Late consequences of bone-marrow infiltration are anaemia,

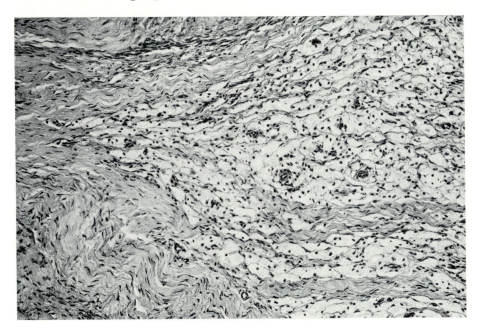

Fig. 36. Xanthomatous accumulation of macrophages in a tendon in primary hyperlipidaemia. Haematoxylin–eosin. × 120.

leucopaenia and thrombocytopaenia. The Gaucher cell is prominent in the enlarged organs, and also may be present in the lungs, kidneys and elsewhere. It is a grossly enlarged macrophage, often over $50\,\mu$ in diameter, occasionally multinucleated, and has a distinctive cytoplasmic appearance. The cytoplasm is pale and exhibits parallel wavy fibrils traversing the otherwise clear cytoplasm in a manner somewhat reminiscent of a spider's web (Figs. 37, 38). The cell carries and stores the enormous quantities of glycolipids (cerebrosides) which are chemically detectable in the affected organs. The major cerebroside contained in the Gaucher cell is behenyl gluco-cerebroside (Rosenberg and Chargaff, 1958) but others have also been demonstrated recently (Yamakawa, 1967). The usual fat stains are negative, but the material is PAS positive. Blood lipid levels are normal, and the exact enzyme abnormality is unknown.

Niemann–Pick disease is a rare condition in which the organs are extensively infiltrated with foamy macrophages, $15–20\,\mu$ in diameter, containing large amounts of cholesterol, phospholipids and a phosphatide called sphingomyelin. The cytoplasm is not fibrillary as in the Gaucher cell, and the intracellular lipids also stain positively with normal fat stains. Many organs, including the brain, are widely affected. Mental retardation and the frequent appearance of cherry-red spots in the ocular

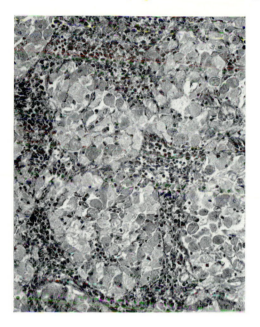

Fig. 37. Clusters of Gaucher cells in red pulp of the spleen
in a child. Azan. × 120.

fundus is a common accompaniment of the condition. The disease is fatal
in infancy. It has been attributed to a deficiency of sphingomyelinase
(Brady, Kaufer, Mock and Fredrickson, 1966).

HISTIOCYTOSIS X

This term is used to encompass a triad of diseases, Hand–Schüller–
Christian disease, Letterer–Siwe disease and eosinophilic granuloma, which
may be varying clinical and pathological manifestations of the same
fundamental process, but which were considered as separate conditions
until about 1940. The term 'histiocytosis X' proposed by Lichtenstein
(1953) seems to be currently favoured, but others have variously re-
commended the terms ideopathic histiocytosis, non-lipoid histiocytosis,
reticulo-endotheliosis, primary reticulo-endothelial granuloma and histiocytic reticulo-endotheliosis.

Letterer–Siwe disease is an acute and fatal disease of infants under one
year of age. It is clinically manifested by a maculo-papular skin rash,
enlargement of lymph nodes, thymus, spleen and liver, and focal lesions
in bones. These features are the result of the infiltration and proliferation
of large pale macrophages containing sparse or absent lipid. Plasma cells,

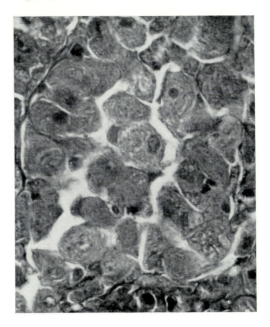

Fig. 38. Gaucher cells showing characteristic fibrillae in cytoplasm. Azan. × 1,000.

lymphocytes, neutrophil polymorphs and eosinophils are also commonly present in variable numbers. Other organs and tissues may be diffusely infiltrated, and pulmonary involvement may lead to honeycomb lung.

Hand–Schüller–Christian disease affects older children and young adults. The syndrome was originally used only in the presence of the triad of exophthalmos, diabetes insipidus and multiple osteolytic defects in the skull bones. It is now evident that this triad is very rare, and that the exophthalmos and diabetes insipidus are consequent upon the skull lesions. Moreover, many cases exhibit lesions in the skin (resembling eosinophilic granulomas) and diffuse histiocytic and eosinophilic infiltrates in the skull and other bones, and in the lungs, spleen, lymph nodes and liver. Although many of the proliferating histiocytes have a 'foamy' appearance and may contain demonstrable lipid, there does not appear to be any underlying disorder of lipid metabolism, and blood cholesterol levels are normal. Multinucleate macrophages (giant cells) are common, together with a prominent admixture of neutrophil polymorphs and eosinophils.

Eosinophilic granuloma of bone is also a disease of children and young adults, and usually presents as a solitary destructive lesion of bone. Microscopy of the lesion reveals a mixture of pale, sometimes foamy,

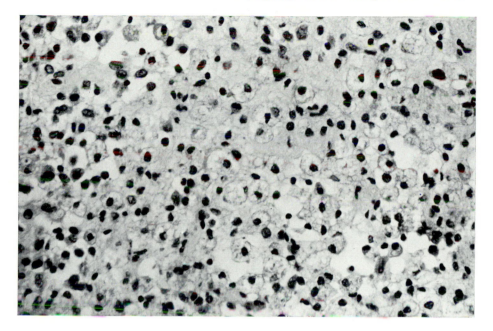

Fig. 39. Sheet of macrophages, some foamy due to lipid content, and scattered
   eosinophils in a solitary lesion of the humerus in Histiocytosis X (eosinophilic
   granuloma type). Haematoxylin–eosin. × 120.

macrophages, eosinophils, and a variable population of fibroblasts,
lymphocytes, plasma cells and neutrophil polymorphs (Fig. 39). The
macrophages form a continuous sheet throughout the lesions with the
other cells present in variable numbers in different areas: where large
numbers of eosinophils are present, small eosinophilic 'micro-abscesses'
may be formed. Multinucleate giant cells are not uncommon, and the
macrophages may contain haemosiderin, fat droplets or phagocytosed
material. Blood eosinophilia is unusual. The disease usually runs a benign
course and solitary lesions are usually cured by surgery or irradiation.
Spontaneous healing has been noted, especially in cases where multiple
lesions have been present (Jaffe and Lichtenstein, 1944). However, deaths
have occurred in young children with multiple lesions who later manifested
visceral lesions (suggesting a transformation to the Hand–Schüller–
Christian type of histiocytosis).

   Intermediate forms of the above conditions may be seen, and there
seem to be very good reasons on clinical and histopathological grounds
for regarding them as being manifestations of the same basic process.
However, it is still useful to equate the eponyms Letterer–Siwe disease
with the acute disseminated form, Hand–Schüller–Christian with the

chronic disseminated form, and eosinophilic granuloma with the benign focal form of the one disease.

TRAUMATIC FAT NECROSIS

Traumatic fat necrosis is a curious focal lesion most commonly encountered in the female breast. The cause of this focal lesion is often obscure, and, despite its name, a history of trauma is often absent. The frequent finding of blood pigments within the lesions has been taken as evidence of previous trauma to the tissues, but the causal relationship is not proven.

Histologically, the affected adipose tissue undergoes a series of changes, first becoming cloudy as the contained neutral fats become saponified; later, the affected fat cells break up, liberating the altered fat which is then taken up by macrophages congregated in the interstices between unaltered fat cells (Fig. 18). These macrophages develop the characteristic finely dispersed granularity of the cytoplasm which results from the uptake of lipid material, and become typical 'lipophages'. It is quite common to observe multinucleate lipophages, particularly in close relation to crystals of fatty acids and cholesterol which may form in the affected area. Calcification and fibrosis eventually ensue. The same pattern of changes may be encountered in bacterial infections of adipose tissues as well as in known chemical and physical injuries to fat. The condition bears no relation to the enzymatic necrosis of fat which occurs in certain forms of pancreatitis accompanied by the release of pancreatic lipases into surrounding tissues.

OLEOGRANULOMAS

This is the term applied to the granulomatous reaction which usually results from the introduction of foreign oily materials into the tissues. They may be seen at injection sites, in lymph nodes after the introduction of oil-containing radio-opaque materials for lymphangiography (Fig. 40), and in the breasts after attempts to change the contours. The lesions characteristically have large rounded spaces containing the foreign material and lined by a layer or two of large foamy macrophages, with similar cells and multinucleate giant cells in the surrounding tissues.

STEROID METABOLISM

Numerous studies on the ability of the adrenal cortex and the liver to perform biosyntheses and biotransformations *in vitro* have been reported. However, until recently, little attention has been given to the particular functions of the various cell types in these organs.

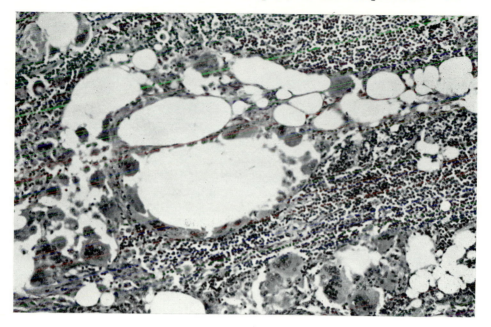

Fig. 40. Pelvic lymph node after lymphangiography. Shows rounded spaces occupied by oily contrast medium with a lining of multinucleate giant cells, and similar cells in adjacent tissues. Haematoxylin–eosin. × 120.

Berliner, Nabors and Dougherty (1964) utilized the phagocytic ability of the sinusoidal macrophages of the adrenal cortex and liver to separate them from the parenchymal cells of these organs They perfused the organs *in vivo* or *in vitro* with iron particles and then minced the tissues. The resulting whole-cell breis were filtered and then suspended in physiological buffer. Separation of macrophages (containing ingested iron particles) from parenchymal cells was achieved by placing the breis in a magnetic field. They then incubated the separated macrophages and parenchymal cells with various steroid substrates. They showed that, in the case of the adrenal cortex, the ratios of 17-hydroxylated steroid products to non-17-hydroxylated products in macrophage (1.07), parenchymal cell (0.283), and macrophage plus parenchymal cell (0.522) incubations indicate that the macrophage is primarily responsible for 17-hydroxylation. The findings also revealed that the capacity for producing corticosterone exists primarily in the parenchymal cell fraction. The zona fasciculata of the adrenal cortex contains more macrophages than the other zones and is mainly responsible for the production of the 'glucocorticoids'; and cortisol (hydrocortisone), an important member of this group, is the principal hormone produced by the adrenal cortex in most species. The hydroxyl group at C-17 of the cortisol molecule is important in enhancing its anti-inflammatory activity (Nicol, Quantock and Vernon-Roberts, 1967), and it has been established that this oxygen function must be introduced into the cortisol molecule prior to those at C-11 and C-21 (Hechter *et al.* 1951; Plager and Samuels, 1954). The data presented by Berliner *et al.* (1964) demonstrate that, in the adrenal cortex, the macrophage is primarily responsible for the 17-hydroxylation of steroids, and that the parenchymal cells have separate distinct functions.

Studies on separated liver cells by Berliner *et al.* (1964) clearly indicated that the hepatic parenchymal cell is the cell in the liver responsible for the conjugation of steroids with glucuronic acid. They also showed that the Kupffer cells are responsible for the reduction of the α,β-unsaturated ketone of ring *A* in corticosteroids; and that this reaction is essential for their conjugation with glucuronic acid by the parenchymal cells in order that they may be excreted as water-soluble glucuronide conjugates.

These functions of adrenal and hepatic macrophages have some common metabolic parameters. It has been shown that phagocytic cells have a very active pentose phosphate metabolic pathway. One consequence of this activity is an increased production of reduced pyridine nucleotides, particularly TPNH, during the phagocytosis of inert particles (Sbarra and Karnovsky, 1959). Reduced pyridine nucleotides are necessary for the introduction of an hydroxyl group at C-17 of progesterone and for the reduction of the α,β-unsaturated ketone in ring *A* of corticosteroids (Plager and Samuels, 1954). In the light of these studies, it is apparent that the macrophage plays an important role in the biotransformation of steroids which, in turn, are important in the regulation of many biological processes including the activity of the macrophage itself (Vernon-Roberts, 1969*b*).

PROTEIN METABOLISM

The presence of abundant rough-surfaced endoplasmic reticulum and free ribosomes within the cytoplasm of macrophages is consistent with the ability of these cells to synthesize proteins. Extensive protein synthesis occurs in macrophages cultivated *in vitro* (Cohn, 1968), but many of the products have not yet been identified. It is also not clear to what extent such synthetic products are retained by the cell or excreted into the environment. However, evidence for the production of some specific biologically active proteins by macrophages has been obtained.

There have been a number of reports concerning the ability of macrophages to synthesize considerable quantities of interferons. Interferons probably play an important role in resistance against virus infections, and appear to act by making cells incapable of replicating viruses by the derepression of antiviral protein. It has been shown that large amounts of

interferons are rapidly produced in response to virus infections of macrophages *in vitro*, and they are also produced 'spontaneously' (in the absence
of virus infections) in cultured macrophages or in response to stimulation
of the cells with bacterial endotoxins (Wagner and Smith, 1968). These
findings have been confirmed by others.

Macrophages are capable of synthesizing at least one component of
the complement complex β1C-globulin (C'3), and also the iron-binding
protein transferrin (Phillips and Thorbecke, 1966; Stecher and Thorbecke,
1967). There is also evidence that macrophages can synthesize prothrom-
bin (Factor 2), proconvertin (Factor 7) and antihaemophilic globulin
(Factor 8) (Slätis, 1958; Gaynor and Spaet, 1966).

AMYLOID FORMATION

Amyloid is a pathological deposit of an abnormal proteinaceous material
which may be found deposited in tissues in a generalized or localized
fashion in a variety of pathological circumstances in both animals and
man. In man, amyloidosis typically affects the kidneys, liver, spleen,
intestine and, to a lesser extent, other organs, but may also be seen
localized to a specific organ. It may occur in the absence of any recognizable predisposing diseases, but is probably more commonly associated
with chronic diseases such as rheumatoid arthritis, carcinoma, osteo-
myelitis, tuberculosis and multiple myeloma. It also occurs in hereditary
form, as in familial Mediterranean fever.

Amyloid is made up primarily of fine fibrils, and chemical analysis
indicates that it is a protein which possibly contains a small amount of
carbohydrate in combination. The amyloid fibril appears to be deposited
in ground substance which may or may not be abnormal (Cohen, 1965). It
is readily recognizable in histological preparations by the demonstration
of metachromasia with crystal violet, birefringence after staining with
Congo red, and fluorescence after staining with Thioflavin T.

It would appear that stimulation of the macrophage system may play a
part in the genesis of amyloid (Cohen, 1965). Among early workers who
studied this aspect of amyloid formation, Smetana (1926) observed that, in
mice stimulated by the injection of killed streptococci, Indian ink accumulated in almost exactly the same sites where amyloid occurs in the kidneys,
spleen, liver, adrenals, and intestines. Moreover, he observed that in mice
given Indian ink and then a series of casein injections, the blockading
effect of Indian ink on the macrophages caused a delay in the appearance
of amyloid. Subsequent studies by Teilum (1952, 1954), also on casein-
induced amyloidosis in mice, indicated that cortisone and nitrogen
mustard administration accelerated the production of amyloid; and this
was interpreted as indicating that inhibition of cellular proliferation of the

'reticulo-endothelial system' is fundamental to amyloid formation. In using the term 'reticulo-endothelial system', Teilum also included reticulum cells, pyroninophilic cells and plasma cells. Teilum propounded a 'two-phase cellular theory of local secretion' in the pathogenesis of amyloid, involving an initial phase of pyroninophilic and plasma cell proliferation accompanied by a rise in serum globulins, followed by the amyloid phase depending on the suppression of proliferating pyroninophilic cells associated with a decrease in globulin levels. In general agreement with this concept Shearing, Comerford and Cohen (1965) found that rabbits treated with casein exhibited a marked increase in the phagocytic activity of the macrophage system (assessed by the carbon clearance technique) which reached a peak at about 20 weeks. During this time none of the animals developed amyloidosis. After 20 weeks, the phagocytic activity declined progressively to normal levels while the animals were still receiving casein, and all the animals were found to have developed amyloidosis during this time.

A number of studies with the electron microscope have demonstrated a close relationship between amyloid fibrils and macrophages. Battaglia (1962) examined the livers of mice with casein-induced amyloid, and was of the opinion that the earliest fibril formation occurred intracellularly within Kupffer cells which accumulated fibrils until rupture of the plasmalemmas ensued and amyloid spilled out of the cells. In similar studies, Sorensen, Heefner and Kirkpatrick (1964) also described amyloid fibrils within Kupffer cells, and some of the fibrils were apparently intimately associated with mitochondria or within the mitochondria. These findings in hepatic amyloid have been confirmed by Cohen (1965), but he has also observed an intimate relationship between amyloid fibrils and cells other than macrophages. Other studies have shown an intimate relationship between amyloid fibrils, and other types of phagocytes, such as mesangial cells in renal amyloid, microglial cells in amyloid of the central nervous system, and the littoral cells lining the sinusoids in amyloidosis of spleen. Despite this microscopic evidence of a close structural relationship between macrophages and amyloid, it does not appear certain whether these appearances represent the local formation of amyloid, fibrils invaginating the cell membrane, or fibrils phagocytosed by the cells. Cohen and his colleagues (Cohen, Gross and Shirahama, 1964) attempted to find the answer to this problem by investigating the ability of macrophages to form amyloid *in vitro*. They cultured explants of spleen from rabbits in which amyloidosis had been induced with casein, and sacrificed the cultures at intervals up to 30 days. Electron microscopy revealed amyloid fibrils in close proximity to macrophage borders and indenting them; they were also disposed intracellularly and occasionally in ergastoplasmic cisterns. In order to determine whether direct formation of amyloid protein might be taking place, tritiated leucine or tritiated tryptophan were introduced into the cultures. Autoradiography subsequently showed that an early cellular uptake of labelled tryptophan was followed by an increase in extracellular label, most of which was located on the amyloid. These findings indicated that amyloid formation can take place in the absence of a blood supply and lend strong support to the concept of local amyloid formation by macrophages. Moreover, they also indicate that amyloid fibrils or their precursors are produced intracellularly.

Cohen (1965) made a detailed review of the constitution and genesis of amyloid and concluded that the genesis of the amyloid fibril would seem most likely related to 'overstimulated reticulo-endothelial cells' (apparently synonymous with macrophages in this context). He was of the opinion that the immunological and other stimuli which stimulate macrophages to produce amyloid, simultaneously induce the proliferation of plasma cells which may be seen also in the lesions of amyloidosis; but it is unlikely that plasma cells are directly involved in the production of amyloid fibrils.

Despite the abundant evidence for a prominent role of the macrophages in the genesis of amyloidosis it appears quite certain that the formation of amyloid is not the exclusive property of these cells, since localized amyloid is frequently and characteristically present within various tumours of the endocrine system such as medullary carcinoma of the thyroid, islet cell tumours of the pancreas and tumours of the parathyroid glands.

Since the time of writing the above, a number of studies have been made of the amino acid sequences in purified amyloid fibril proteins. They indicate that the amyloid fibril proteins are derived from homogeneous immunoglobulin light chains of the variable region subgroup V_{kl}. If these findings reflect the situation in all types of amyloidosis, they establish a strong relationship between amyloid deposition and a disturbance of immunoglobulin synthesis. Since there is universal agreement that macrophages do not possess the capacity for synthesizing immunoglobulins, the findings indicate that the presence of amyloid fibrils in intimate extracellular and intracellular relationship with macrophages is the result of deposition followed by the endocytosis of the fibrils by macrophages and is not the result of their synthesis by these cells. Recent evidence has also confirmed that macrophages actively ingest amyloid fibrils.

6

MACROPHAGES AND CELL-MEDIATED IMMUNITY

ACQUIRED CELLULAR RESISTANCE TO INFECTION

Natural resistance to infection generally refers to the resistance possessed by animals which have not been actively or passively immunized against a particular micro-organism, and is sometimes referred to as non-specific resistance. The relative importance of the various factors which may play a part in non-specific resistance vary according to the nature and dose of the infecting micro-organism. While there are many examples of species and strain differences in susceptibility to infection in animals, in the individual such factors as genetic susceptibility, age, the integrity of mechanical barriers such as skin and mucous membranes, the bactericidal activity of various body secretions (tears, sweat, gastric secretions, etc.) and the influence of hormones on macrophage function may all play a part. Moreover, naturally-occurring antibody and complement play an important role in the phagocytosis and destruction of micro-organisms by macrophages and polymorphonuclear leucocytes under conditions where active or passive immunization has not taken place.

Acquired resistance refers to resistance manifested by animals which have actively mounted an immune response against a micro-organism's components or products as a result of previous exposure or deliberate immunization, or which have been passively immunized by the receipt of serum or cells from an immunized animal. Acquired antibacterial immunity is, in many instances, related to circulating antibody which displays a high degree of specificity and which enhances the phagocytosis and intracellular killing of bacteria by macrophages (see Chapter 3). In some cases, particularly when the infecting organism is capable of living and multiplying within macrophages (i.e. the organism is a facultative intracellular parasite), circulating antibody does not play a significant part, and the enhanced bacterial killing and resistance to re-infection is probably mediated by antibody bound to the surfaces of cells. This latter type of cell-mediated immunity is less specific than when circulating antibody is involved and usually confers increased resistance against other facultative intracellular parasites.

Acquired cellular resistance probably plays the predominant role in defence against infection by *Mycobacteria*, *Brucella*, *Listeria* and *Salmonella*, although humoral immunity also plays a part in resistance to infections with *Salmonella*. There is also evidence that acquired cellular resistance plays a part in defence against infection with such varied organisms as *Histoplasma*, *Candida*, *Plasmodium*, *Trypansoma*, *Leishmania*, *Toxoplasma* and, possibly, *Corynebacteria*.

NATURE AND SPECIFICITY OF ACQUIRED CELLULAR RESISTANCE

The increased capacity of macrophages from immunized animals to destroy or suppress the multiplication of micro-organisms has been widely demonstrated. Acquired antibacterial immunity relates almost exclusively to those organisms which can survive and multiply within macrophages, and generally exhibits some or all of the following characteristics: (1) resistance cannot be induced by killed vaccines but follows infection with avirulent organisms or recovery from the disease; (2) resistance is associated with delayed-type hypersensitivity towards the corresponding microbial antigen; (3) resistance is non-specific; (4) macrophages from resistant animals have an increased capacity to destroy ingested micro-organisms in the absence of antibody; and (5) resistance cannot be passively transferred to normal animals by means of serum, but may in some circumstances be transferred by means of cells from immunized animals. Tuberculosis is somewhat exceptional in that dead tubercle bacilli are active both as immunizing and sensitizing agents, and almost certainly this relates to the capacity of mycobacteria to act as adjuvants for the induction of delayed-type hypersensitivity to protein antigens in general (Mackaness, 1967). Apart from tuberculosis, in the other typical intracellular infections, such as brucellosis, listeriosis and salmonellosis, the killed organisms do not produce sensitization and give little or no protection against subsequent infection. This lack of effect of killed organisms in producing sensitization is in striking contrast to the enhanced microbicidal activity which is acquired during active infection with virulent or attenuated strains of the same organisms (Mackaness and Blanden, 1966). In all cases, the macrophages of actively infected animals exhibit greatly enhanced ability to kill the infecting organism, but it is also notable that animals infected with one species of intracellular parasite become partially or completely resistant to infection with other unrelated species; and their macrophages exhibit similar cross-resistance to infection *in vitro*. Consistent with these findings, acquired cellular resistance does not significantly involve humoral antibacterial antibody. Instead, it appears to depend on the ability of the host to acquire a population of host macrophages with enhanced microbicidal activity (Mackaness and Blanden, 1966). However, despite the lack of specificity, there is little doubt that acquired cellular resistance is dependent upon immunological processes, since enhanced antibacterial activity cannot be detected in the host or its macrophages until the tissues have had time to respond immunologically, and convalescent animals gradually lose their resistance but re-acquire it at an accelerated rate during re-infection (Mackaness, 1967). It is thus apparent that both specific sensitization of the tissues and sustained antigenic stimulation are required for the induction of acquired cellular immunity.

The macrophages of animals recently recovered from a sublethal infection by one of the intracellular bacterial parasites are found to differ from normal cells in their morphology and are also capable of ingesting and destroying a wide variety of organisms: macrophages with these properties are said to be 'activated'.

MECHANISM AND MORPHOLOGY OF MACROPHAGE ACTIVATION

It has been shown that in mice infected with either *L. monocytogenes* or BCG there is an intense proliferation of lymphoid cells in the spleen, a coincident proliferation of resident macrophages in the peritoneal cavity, and the subsequent emergence of a population of macrophages with increased ability to phagocytose inert particles and to spread on a foreign surface (North, 1969*a*). The division of both lymphoid cells and macrophages appeared to be essential for the production of the activated macrophages. The temporal relationship between cellular proliferation and the production of activated macrophages remains unchanged although the time between infection and the onset of these events depends upon the individual infecting organism used. North (1969*b*) has shown that the fixed macrophages of the liver (Kupffer cells) undergo mitosis in large numbers during an infection with *L. monocytogenes*, and this always precedes the expression of efficient host immunity to infection. In addition to their increased spreading ability and phagocytic activity, activated macrophages also exhibit increased respiration rates, hydrolytic enzymes, microbicidal activity and cytopeptic activity (see Mackaness, 1970*a*). Morphologically, the activated macrophages are larger than normal macrophages, and have increased numbers of cytoplasmic organelles, especially lysosomes, and pinocytotic vesicles. It is apparent that the increased functional activity of the cells may play a major part in the ability of the activated macrophages obtained from animals sensitized to one organism, to kill other antigenically-unrelated facultative intracellular parasites in the absence of specific opsonins.

It has been shown that the specific cellular resistance and delayed-type hypersensitivity to *L. monocytogenes* can be successfully transferred from immune (convalescent) mice to normal mice by means of splenic lymphoid cells or thoracic duct lymph (Mackaness, 1970*a*). In contrast, the transfer of serum from immune animals does not confer any protection. It seems reasonably certain that the numbers of lymphoid cells transferred in these experiments were unlikely themselves to give rise to a large population of offspring macrophages, but nevertheless they were able to induce increased immunity in the recipients of a type which is solely expressed through activation of recipient macrophages. It thus seems that the transfer and specific recall of acquired cellular immunity is mediated by immuno-logically-committed lymphocytes, and this is also indicated by the observation that antilymphocyte globulin blocks the transfer of cellular immunity and delayed-type hypersensitivity by immune spleen cells (Mackaness and Hill, 1969). During mycobacterial infection with BCG, mice also become highly resistant to *Listeria* (Blanden, Lefford and Mackaness, 1969). The fact that the lymphoid cells of BCG-infected mice can transfer tuberculin sensitivity to normal animals, but do not give enhanced resistance against *Listeria*, indicates that delayed-type hypersensitivity is due to the presence of committed lymphocytes whereas enhanced antibacterial resistance is due to the presence of activated macrophages. The administration of a small dose of BCG to mice previously given lymphoid cells from BCG-infected donors causes activation of macrophages in the recipient animals accompanied by increased resistance to *Listeria* challenge (Mackaness, 1969). These findings indicate that the macrophages of animals with delayed-type hypersensitivity are influenced by sensitized lymphoid cells in the presence of antigen. In the presence of antigen, it is possible that sensitized lymphoid cells could produce a substance which then causes the activation of the macrophages, or could produce antibody cytophilic for macrophages which could then in turn react directly with the antigen to become activated. Both possibilities merit further examination.

ROLE OF ANTIGENIC STIMULATION OF SENSITIZED LYMPHOID CELLS

It has been shown that lymphocytes, when stimulated *in vitro* by specific antigen, release into the medium a soluble material which inhibits the migration of normal macrophages (not from sensitized donors) from capillary tubes. This migration inhibitory factor (MIF) is only produced in response to the specific sensitizing antigen, and is not produced by the cells in the absence of antigen (David *et al.* 1964) (Figs. 43, 44). It now seems quite certain that MIF is not an immunoglobulin.

Other factors in addition to MIF are also produced by sensitized lymphocytes in the presence of specific antigen (see David, 1970). They include the immunoglobulins, a factor chemotactic for macrophages, a factor which causes erythema and inflammation in guinea-pig skin, factors cytotoxic to cells in culture, factors which cause blastogenesis in non-sensitized lymphocytes, and factors with interferon-like activity. While these findings indicate that sensitized lymphocytes produce a variety of factors in response to antigenic stimulation, and that some of these factors have effects on normal macrophages, a factor stimulating the activation of macrophages has not yet been demonstrated *in vivo* or *in vitro*. However, the chemotactic factors and MIF could be envisaged as functioning in harmony *in vivo*, firstly to attract macrophages into areas of antigen (microbial) challenge, and secondly to retain them at the same site to become activated and deal more effectively with the invaders. In this connection, it has been observed that the subcutaneous injection of specific antigen into sensitized animals causes immobilization and clumping of peritoneal macrophages (Nelson, 1969). Moreover, in response to re-injection of specific antigen there is a marked increase in mitotic activity in peritoneal macrophages of BSA- or *Mycobacteria*-sensitized mice (Forbes, 1965), and in the alveolar macrophages of *Mycobacteria*-sensitized rabbits (Myrvik, Leake and Oshima, 1962).

ROLE OF CYTOPHILIC ANTIBODIES

There is evidence in favour of and against the idea of a role for macrophage cytophilic antibody initiating some of the manifestations of delayed-type hypersensitivity (Nelson, 1969), and acquired cellular resistance is generally associated with the state of delayed-type hypersensitivity. Apart from Salmonella infections, where humoral immunity also plays a part in the resistance to infection and cytophilic antibodies have been found in the serum of infected mice (Rowley, Turner and Jenkin, 1964), cytophilic antibodies have not been demonstrated in the serum during infection with *Mycobacteria*, *Brucella* or *Listeria*. Moreover, if cytophilic antibodies were present within the serum, then this would imply that serum from infected or convalescent animals should be capable of transferring macrophage reactivity with antigen to induce macrophage activation, but this is not the case. The failure to demonstrate circulating cytophilic antibodies and the failure of serum to transfer sensitization could be explained by the antibodies being so strongly cytophilic that they are removed from the serum by becoming bound to the surfaces of the cells as fast as they are produced. The presence of cytophilic antibody would enable the macrophages to react directly to contact with antigen, and would help to explain the increased DNA synthesis and acid hydrolase content of macrophages from convalescent animals exposed to specific antigen *in vitro*, the fact that trypsin-treated macrophages are not immobilized by antigen in the presence of sensitized lymphoid cells (David, Lawrence and Thomas, 1964), the necessary participation of blood monocytes in skin reactions in delayed-type hypersensitivity (Mackaness, 1970*b*), and other reactions in the macrophages of infected or re-infected animals.

THE ROLE OF MACROPHAGES IN DELAYED-TYPE HYPERSENSITIVITY

Delayed-type hypersensitivity is the term used to describe the state of animals during which contact with antigen applied to the skin by means of intracutaneous injection, scratch test, patch test or by painting (depending upon the nature of the antigen) results in the tardily-occurring onset of skin reaction which reaches a maximum 24–48 hours after antigen application. The reaction involves local oedema and erythema, and is accompanied by a cellular infiltrate in the underlying tissues which is largely composed of monocytes and macrophages (Figs. 41, 42). In severe reactions, blistering or necrosis of the skin may occur. These changes are accompanied by the proliferation of immunoblasts in lymph nodes draining the area (Turk, 1967).

Delayed-type hypersensitivity reactions probably accompany most, if not all, acute and chronic bacterial infections such as tuberculosis, brucellosis and tularaemia; on revaccination with vaccinia, during lymphogranuloma venereum, and other virus infections; during fungal infections such as histoplasmosis, coccidioidomycosis and candidiasis; during protozoan infections such as malaria, toxoplasmosis and leishmaniasis; and during metazoan infections such as trichinosis and ascariasis. The reaction is elicited only by the specific antigen or antigen with closely similar structure, and arises within about 5 days after infection or somewhat longer in tuberculosis.

Although the delayed-type hypersensitivity skin reaction can be elicited in sensitized animals by the injection of pure antigen, it is unusual for the antigen alone to be capable of inducing the hypersensitive state in the normal animal; and it is usually necessary to administer pure protein antigens emulsified in Freund's complete adjuvant to induce successfully the state of delayed-type hypersensitivity. This may well be related to the content of *Mycobacteria* in Freund's complete adjuvant, although the incomplete form of adjuvant, lacking *Mycobacteria*, and other techniques are sometimes successful in producing a form of delayed-type hypersensitivity with some antigens.

Delayed-type hypersensitivity (contact sensitivity) to certain chemicals, such as formalin, picryl chloride, dinitrofluorobenzene and oxazolone, follows the application of these substances to the skin by painting, so that

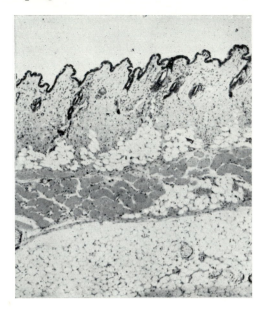

Fig. 41. Normal guinea-pig skin. Haematoxylin–eosin. × 35.

re-application is followed by a temporally and morphologically similar skin reaction to that seen with the pure protein antigens. These substances are all capable of firm binding to proteins, and their capacity for binding to proteins in the skin almost certainly underlies the sensitization which follows the formation of haptens. Since the protein fraction of the hapten is contributed by the host, the sensitization in contact sensitivity is directed against the incorporated determinant group. It seems likely that a similar process is involved in chronic eczematous eruptions which result from contact with various chemicals which probably combine with skin proteins to confer foreign specificity on them.

Delayed-type hypersensitivity also accompanies the manifestations of transplantation and tumour immunity, and is present in some forms of auto-immune phenomena.

CELLS PARTICIPATING IN DELAYED-TYPE HYPERSENSITIVITY REACTIONS IN THE SKIN

In the skin reactions of delayed-type hypersensitivity to soluble or particulate antigens, macrophages are generally present in abundance, although lymphocytes may be present in considerable numbers also. Reports on the relative proportions of the two types of cell, the macrophage (or monocyte) and the lymphocyte, show considerable apparent lack of agreement; and

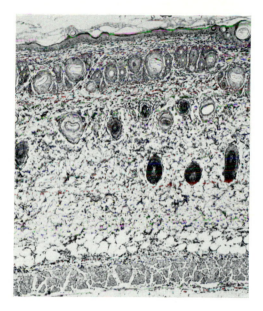

Fig. 42. Delayed-type hypersensitivity reaction showing oedema and heavy infiltration by mononuclear cells in dermis 48 hours after intradermal injection of tuberculin in guinea-pig previously sensitized with Freund's complete adjuvant. Haematoxylin–eosin. × 35.

this could reflect differences in the time of assessment after antigen exposure, the type of antigen, the type of animal, and differences in terminology with regard to cell morphology. Although the estimates of the proportion of macrophages varies from 10 to 100 per cent, from the most reliable studies it would seem that the true figure is in the range of 50 to nearly 100 per cent. Turk and his colleagues reported that in peak reactions to tuberculin in guinea-pigs, about half of the cells in the skin lesions were macrophages; whereas in peak reactions of contact sensitivity, most of the cells were lymphocytes (Turk, Heather and Diengdoh, 1966; Turk, Rudner and Heather, 1966). In these studies, macrophages were identified as cells exhibiting high levels of acid phosphatase activity. This distinction may not be completely reliable since it is known that monocytes may have low or absent levels of acid phosphatase; moreover, up to 90 per cent of small lymphocytes and most of the immunoblasts in draining lymph nodes during the development of contact sensitivity to oxazolone may contain lysosomes exhibiting acid phosphatase activity (Diengdoh and Turk, 1965). Extensive studies have been carried out by Spector and his colleagues on the identity and kinetics of cells participating in delayed-type hypersensitivity (Spector, 1967). They found that in delayed-type hypersensitivity to intradermal tuberculin, and in contact sensitivity to DNCB or picryl chloride, the histology of the reactions was consistent with an inflammatory reaction of exaggerated intensity but delayed onset, and that at the peak of the reactions the inflammatory exudate was predominantly composed of mononuclear phagocytes, either monocytes or mature macrophages. Labelling studies, wherein serial examinations of blood and cellular exudate were carried out after a single pulse of tritiated thymidine given well beforehand, revealed that an identical proportion (50 per cent) of circulating monocytes and exudate mononuclear cells were labelled at 24 hours after tuberculin challenge.

There is little participation of circulating small lymphocytes in skin reactions to tuberculin (Spector, 1967) although large lymphocytes have been reported to be present in significant numbers (Kosunen, Waksman, Flax and Tihen, 1963). The majority of the latter cells are probably monocytes and immature macrophages, as evidenced by electron microscopic examination of the lesions. There is thus little doubt that, as in sterile non-immunological inflammatory lesions, the majority of the cells which appear in delayed-type hypersensitivity reactions have the morphology and labelling characteristics of blood monocytes, and are mostly recently derived cells which have arisen by mitotic division at some site other than in the reaction site and migrate there via the circulation. Once having emigrated into the reaction site, many of the monocytes transform into mature macrophages, and in established reaction sites only about 4 per cent of the cells exhibit DNA synthesis, as indicated by the uptake of tritiated thymidine. In chronic granulomata associated with delayed-type hypersensitivity, there is a sustained emigration of monocytes into the reaction site, but most of the new cells appearing arise by cellular proliferation within the lesion itself (Spector, 1967). The importance of monocytes in delayed-type hypersensitivity is further exemplified by the fact that delayed-type hypersensitivity can be abolished by doses of whole-body X-irradiation which eliminate monocytes from the circulation (Volkman and Collins, 1968).

THE PERITONEAL MACROPHAGE DISAPPEARANCE REACTION

Macrophages are involved in a reaction which is seen only in hypersensitive animals. It has been found that the subcutaneous injection of tuberculin (PPD) into hypersensitive (BCG-vaccinated) guinea-pigs is followed by the almost complete disappearance of macrophages from the fluid phase of peritoneal exudates (Nelson and Boyden, 1963). The reaction is immunologically specific and cannot be elicited in unsensitized animals, nor can it be elicited in BCG-vaccinated animals by antigens other than tuberculin. However, the reaction can be elicited by bacterial endotoxins, but to the same extent in sensitized and normal animals. Studies on the fate of the macrophages disappearing from the peritoneal fluid (Nelson, 1969) have shown that the number of cells 'lost' can be completely accounted for by the cells sticking together and to the lining of the peritoneal cavity. The number of macrophages is reduced for about two days after antigen injection. On the fourth day there is a significant rise in the number of lymphocytes in the peritoneal cavity, and this is accompanied by the appearance of 2–4 per cent of blast-type cells, some of which are in mitosis.

It is of interest that the peritoneal macrophage disappearance reaction can be passively transferred with whole peritoneal or alveolar cell populations or with peritoneal lymphocytes, but not with lymph node lymphocytes. This suggests that transferred reactivity may be mediated by macrophages carrying cytophilic antibody, or by lymphocytes synthesizing cytophilic antibody which becomes attached to recipient's macrophages; or by sensitized lymphocytes synthesizing a pharmacological mediator similar to MIF. The fact that reactivity cannot be transferred by means of serum alone but can be transferred by normal cells treated with serum from a hypersensitive animal, favours the concept that reactivity is conferred by cytophilic antibody bound to macrophages, since efforts to demonstrate cytophilic antibody capable of binding to lymphocytes in the guinea-pig have not been successful (Nelson, 1969).

The peritoneal macrophage disappearance reaction is inhibited by pretreatment of animals with the anticoagulants heparin and sodium warfarin, but not by cortisone, promethazine or reserpine. However, it appears unlikely that clotting is involved in the clumping of macrophages observed during the reaction since electron microscopic studies have not revealed any evidence of fibrin formation (Nelson, 1969).

INHIBITION OF MACROPHAGE MIGRATION

There have been many studies on the migration of macrophages, obtained from animals with delayed-type hypersensitivity, after exposure to specific antigen. Macrophages in suspension, in monolayers, or in explants of lymphoid tissues have all been examined, and various authors have reported that macrophages from tuberculin-sensitive guinea-pigs may exhibit a variety of morphological and functional changes upon exposure to tuberculin. The effects observed include the inhibition of the normal migration of macrophages, the transformation of small mononuclear cells into macrophages, the transformation of 'lymphocytes' into lysosome-rich proliferating cells, and DNA synthesis in macrophages

Fig. 43. 24-hour migration from capillary tube *in vitro* of macrophages of the peritoneal cell population harvested from a non-sensitized guinea-pig. PPD (30 μg/ml) is present in the culture supernatant.

(see Dumonde, 1967). These events may lead to overcrowding and cell detachment, and the establishment of confluence in monolayer cultures.

A more precise and semi-quantitative assessment of the *in vitro* effects of antigen on macrophage migration was introduced by George and Vaughan (1962) and has been widely investigated by David and his colleagues (see review by David, 1970). In this technique, peritoneal exudate cells are packed by centrifugation into capillary tubes sealed at one end, and the tubes are transected at the cell–supernatant interface. The portion of tube which contains the packed cells is then placed in a chamber containing tissue culture medium with serum added. The chambers are incubated at 37 °C for 24–48 hours during which time the macrophages normally migrate out of the tubes in a fan-like pattern (Fig. 43). When the exudate cells are obtained from animals with delayed-type hypersensitivity and are incubated in the presence of the sensitizing antigen, the migration of the cells is consistently inhibited (Fig. 44). The degree of inhibition can be measured by assessing the area over which the cells have migrated, and comparing it with controls in which antigen has not been added. This is a specific immunological reaction which occurs with protein antigens, hapten–protein conjugates and with synthetic antigens.

In early studies by David and his colleagues, migration inhibition was not produced by incubating normal cells with serum from hypersensitive animals in the presence of the sensitizing antigen; nor was prior incubation of normal cells with serum from hypersensitive animals successful in

Fig. 44. Inhibition of 24-hour migration of macrophages of the peritoneal cell population harvested from a guinea-pig previously sensitized with Freund's complete adjuvant. PPD (30 μg/ml) is present in the culture supernatant.

conferring susceptibility to the sensitizing antigen in subsequent culture. However, Dumonde (1967) has reported that the migration of guinea-pig macrophages passively sensitized *in vitro* with γ_2-cytophilic antibody to sheep erythrocytes is inhibited by sheep erythrocytes, though not by soluble erythrocyte protein. In contrast, macrophages from guinea-pigs immunized with, and displaying delayed-type hypersensitivity to, sheep erythrocytes, exhibit inhibition of migration in the presence of both sheep erythrocytes and soluble erythrocyte protein. Similarly, others have reported that normal guinea-pig macrophages incubated with serum containing cytophilic antibodies to PPD subsequently exhibit inhibition of migration in the presence of PPD but not in the presence of β-lactoglobulin, whereas normal macrophages incubated with serum containing cytophilic antibodies to β-lactoglobulin exhibit inhibition of migration in the presence of β-lactoglobulin but not with PPD (Amos, Gurner, Olds and Coombs, 1967; Heise and Weiser, 1967). As David (1970) has pointed out, it seems likely that these observations are probably due to the formation of antigen–antibody complexes which can themselves inhibit the migration of normal cells when formed in antibody excess. Moreover, the inhibition of macrophage migration in the presence of immune complexes is unrelated to the events taking place when the migration of cells from hypersensitive animals is inhibited in the presence of the sensitizing antigen.

Peritoneal exudate cells are a heterogeneous population, the majority of cells being macrophages in various stages of maturity but with a small number of lymphocytes also consistently present; and the role of the individual cell types is obviously fundamental to the understanding of migration inhibition. David and his co-workers found that, in admixtures of normal and sensitive cells, the number of specifically sensitive peritoneal exudate cells can be reduced to as little as $2\frac{1}{2}$ per cent of the total population, and still the migration of the entire population is inhibited by antigen. This necessitated that the sensitized cells were viable and should be closely packed with the normal cells in the capillary tubes, since no migration inhibition of the normal cells was observed when the normal and sensitive cells were cultured separately in capillary tubes within the same chamber. Further studies have shown that sensitized spleen or lymph node lymphocytes are not themselves inhibited from migrating *in vitro* by antigen, but when mixed with normal macrophages from unsensitized animals the migration of the entire population is inhibited by antigen (Dumonde, 1967; David, 1970).

The separation of peritoneal exudates from hypersensitive animals into separate lymphocyte and macrophage fractions has further elucidated the interaction of these cells in migration inhibition. In the absence of lymphocytes, the macrophage fraction is no longer inhibited by the sensitizing antigen; but when the lymphocytes are added back to the macrophages, or added to unsensitized peritoneal exudate cells in the proportion of as little as 2 per cent, the resulting population is inhibited by antigen (Bloom and Bennett, 1966; Dumonde, 1967). From these observations, it seems clear that there is an initial interaction between antigen and sensitized lymphocytes which is subsequently followed by the inhibition of migration of macrophages which need not necessarily be obtained from animals with delayed-type hypersensitivity. It is also clear that, in the presence of antigen, one sensitized lymphocyte can effectively inhibit several thousand macrophages, since as little as 2 per cent of added peritoneal lymphocytes are effective and it is unlikely that all these cells are sensitive.

It is now known that, after contact with specific antigen, sensitized lymphocytes *in vitro* release a soluble factor into the supernatant which inhibits the migration of normal macrophages (Bloom and Bennett, 1966; David, 1966). This factor has been called migration inhibition factor (MIF). MIF is released into the supernatant after about 6 hours of incubation of sensitized lymphocytes with antigen, and is not an immunoglobulin, but its exact chemical nature is hitherto undetermined. It is resistant to RNAse and DNAse, but its production is blocked by puromycin or mitomycin.

From the above studies on migration inhibition, it would appear that in some of the manifestations of delayed-type hypersensitivity, at least *in vitro*, macrophages are passive participants and that the sensitized lymphocyte is the specifically reactive cell. However, the fact that the macrophages alone are inactivated as a result of the interaction of antigen with sensitized lymphocytes, indicates that some of the manifestations of delayed-type hypersensitivity *in vivo* may be an expression of an interaction between lymphocytes and macrophages. In addition to MIF, other factors are produced by sensitized lymphocytes in the presence of specific antigen *in vitro* (David, 1970), and among these are factors which are chemotactic for macrophages and factors causing erythema and inflammation in guinea-pig skin. Thus the interaction of antigen with sensitized lymphocytes could cause the liberation of factors which, if present at a skin site, could be envisaged as producing many of the local changes which are observed in the skin reactions of delayed-type hypersensitivity. Lymphocytes are constantly present in the skin lesions of delayed-type hypersensitivity, but it seems that this is a non-specific migration in that sensitized and non-sensitized cells are both present. Moreover there is evidence that macrophages may themselves play a role in the initiation of the delayed-type hypersensitivity reaction in the skin by virtue of possessing cytophilic antibodies which enable them to react specifically to antigen.

ROLE OF MACROPHAGE CYTOPHILIC ANTIBODIES IN DELAYED-TYPE HYPERSENSITIVITY

If cytophilic antibodies are involved in delayed-type hypersensitivity reactions, it should be possible passively to transfer reactivity to antigen to normal animals by means of serum from hypersensitive animals. In experiments on contact sensitivity to oxazolone in the mouse, Asherson and Zembala (1970) found that purified populations of peritoneal macrophages and purified peritoneal lymphocytes both transfer contact sensitivity, but the transfers differed in their time course. The reactions to peritoneal lymphocytes were greater at 24 hours than at 12 hours, whereas the reactions transferred by peritoneal macrophages were greatest at 12 hours. This suggested that different mechanisms may be involved in the transfer of sensitivity by the two types of cells. Serum from mice immunized with oxazolone, or normal macrophages incubated with serum from immunized mice and then washed, were also capable of transferring contact sensitivity. Others have shown that skin reactions to PPD, having the features of delayed-type hypersensitivity, can be transferred using purified peritoneal macrophages (Turk and Polak, 1967) or serum (Dupuy, Percy and Good, 1969). These findings indicate that macrophages which have acquired a serum factor (presumably cytophilic antibody) can transfer skin reactions. Additional evidence in support of a role for cytophilic antibody in reactions associated with delayed-type hypersensitivity is provided by the findings that the peritoneal macrophage disappearance reaction can be transferred by means of peritoneal cells carrying a cytophilic antibody (Nelson, 1969), and that the susceptibility of macrophages to MIF can be abolished by treatment of sensitive cells with trypsin (David, 1970).

Against the concept of a role for macrophage cytophilic antibodies is the fact that no correlation has been found in guinea-pigs between cytophilic antibodies and delayed skin reactivity, and attempts to transfer skin reactivity by means of serum from hypersensitive donors or by normal peritoneal cells incubated with such serum has not been uniformly successful; and perhaps more significantly it has been shown that normal guinea-pigs injected with serum containing 7S γ_2-globulin cytophilic antibodies do not exhibit the manifestations of delayed-type hypersensitivity despite the presence of circulating cytophilic antibody (Nelson and Boyden, 1967).

It is evident that no distinct conclusions can be drawn about the role of cytophilic antibodies in delayed-type hypersensitivity. The various manifestations of delayed-type hypersensitivity may be mediated by different mechanisms and it is conceivable that macrophage cytophilic antibodies may play a part in only some of these reactions, such as the peritoneal macrophage disappearance reaction and contact sensitivity. Finally, it is by no means certain that any or all passively transferred reactions are true examples of delayed-type hypersensitivity.

THE ROLE OF MACROPHAGES IN TRANSPLANTATION IMMUNITY AND TUMOUR IMMUNITY

It is apparent that macrophages are prominent participants in only some of the manifestations of those immunological reactions which lead to the rejection of allografts, tumour immunity and graft-versus-host reactions. In discussing the role of macrophages in these complex phenomena, the following terminology is based upon that proposed by Gorer, Loutit and Micklem (1961): *autograft, autologous graft* or *autochthonous graft* for cells or tissues of the grafted animal itself; *isograft, isologous graft, isogenic graft*, or *syngeneic graft* for cells or tissues grafted between genetically identical twins or animals of the same inbred strain; *homograft, homologous graft* or *allogeneic graft* for cells or tissues grafted between genetically different animals of the same species; and *heterograft, xenograft, heterologous graft* or *xenogeneic graft* for cells or tissues grafted between animals of different species.

MORPHOLOGICAL REACTIONS TO ALLOGENEIC TISSUES AND CELLS

In the first few days following the primary application of a skin allograft, the allograft elicits the same response as an autograft in that it becomes attached to the recipient by granulation tissue and subsequently becomes vascularized. By 5–7 days after grafting, mononuclear cells appear in the perivascular regions around capillaries and venules in the dermis. At 7–8 days the capillaries and venules of the dermis exhibit loss of endothelium and disintegration of their walls, and the walls of small arteries and veins within the graft become infiltrated by mononuclear cells, accompanied by swelling of the endothelium and obstruction of the lumen by fibrin and cells. The damage to the walls of blood vessels and the obstruction to blood flow through them results in ischaemic necrosis of the graft, which is first of a patchy nature affecting the epidermis alone, but later involves the dermis. The death of the graft, which occurs about 10–12 days after grafting, is accompanied by a heavy infiltration of the graft and its bed by neutrophil polymorphs, macrophages and mature plasma cells. The majority of the mononuclear cells which appear initially in the rejection process are large lymphoid cells having a large single indented nucleus with a prominent nucleolus, prominent Golgi apparatus, numerous free ribosomes, little smooth endoplasmic reticulum, few mitochondria and absent rough endoplasmic reticulum. These cells, which are generally similar to the pyroninophilic 'immunoblasts' seen in lymph nodes and spleen during immunological responses (see Chapter 7), are often also called 'immunoblasts' although this does not necessarily imply that these cells have an identical function during immunization and graft rejection. The cells infiltrating grafts are also sometimes called 'haemocytoblasts', 'large pyroninophilic cells', 'lymphoblasts' or 'transitional cells'. Other cells forming the early infiltrate in rejecting skin allografts have the morphology of normal small and medium lymphocytes; and cells having the morphology of monocytes and macrophages have been reported to make up a large proportion of the infiltrating cells in skin allografts in guinea-pigs, mice and rats, but not in rabbits.

Organ allografts, as exemplified by renal transplants, differ from skin allografts in that the transplanted organ is perfused by the blood of the recipient from the time when the vascular anastomoses are surgically established. In the case of renal allografts, 'immunoblasts', similar to those seen in the early infiltration into skin allografts, are seen in the peritubular capillaries and venules of the cortex and outer medulla about three days after transplantation. Labelling studies have shown that these cells are of host origin (Porter and Calne, 1960). Also present at this stage are macrophages, which possess more RNA than is usual, mainly in the form of free ribosomes (Porter, 1967). The infiltrating cells lie close to the endothelial cells lining peritubular blood vessels, and subsequent cytoplasmic fusion of endothelial and infiltrating cells (Kountz *et al.* 1963) is followed by swelling and separation of endothelial cells and disruption of the peritubular capillary walls. Some of the infiltrating immunoblasts undergo division within the graft tissues. During the following days, there is progressive destruction of peritubular capillaries and venules, with immunoblasts, macrophages, and increasing numbers of plasma cells infiltrating the graft. The leakage of fluid and cells from the damaged vessels into the interstitial tissues causes swelling of the kidney and inhibits blood flow through the organ so that urinary output falls and necrosis of the proximal tubules occurs. Finally, macrophages and neutrophil polymorphs infiltrate the organ in increasing numbers, there is fibrinoid necrosis of the walls of afferent arterioles and interlobular arteries which become occluded by thrombus and infiltrating cells, and the graft undergoes ischaemic necrosis. A similar sequence of cellular infiltration and vascular changes is seen during the rejection of allografts of liver, heart, lungs and spleen.

If the recipient animal is again grafted with skin or an organ from the same donor animal which contributed the first graft, the second graft is destroyed more rapidly than the first graft. This accelerated rejection phenomenon which occurs if 2–3 weeks elapse between first and second graft application is known as the 'second-set' reaction, and is due to sensitization of the host by the first graft. During second-set reactions, skin allografts exhibit heavy infiltration with immunoblasts 2–3 days after grafting, and death of the graft occurs over a period of 5–7 days. If the interval between application of first and second grafts is shortened to within 14 days, very little vascularization takes place and results in a pale, apparently avascular, graft which rapidly degenerates. This is known as the 'white-graft' reaction, and histologically the bed of the graft exhibits a heavy infiltration with polymorphs and has been likened to an Arthus reaction.

Studies on the cellular changes in lymph nodes and spleen after transplantation of allogeneic skin or organ grafts have shown generally similar changes in all cases. Within 2–4 days after primary allografting, pyroninophilic immunoblasts appear amongst the small and medium lymphocytes of the mid and deep cortex of draining lymph nodes especially around post-capillary venules in thymus-dependent areas (Scothorne and McGregor, 1955; Parrott, 1967). These blast cells are identical to those which appear in the graft itself. They increase in numbers to a maximum about 8–10 days after grafting, at which time macrophages containing ingested material are also present in increased numbers, but thereafter progressively decrease in number at the same time as the germinal centres appear and increase in size (Parrott, 1967). Smaller numbers of pyroninophilic immunoblasts are present in distant lymph nodes and spleen over the same period as these cells are present in draining nodes. The changes in the spleen predominate after the intravenous injection of allogeneic cells. The draining nodes are usually morphologically normal about 4 weeks after grafting. During second set reactions, the changes observed are generally similar to those in the first set reaction, but appear earlier and with greater intensity. The exact origin of the immunoblasts which appear in lymph nodes after skin grafting is still uncertain, but the fact that neonatally thymectomized mice are unable to reject skin allografts and do not exhibit pyroninophilic immunoblast proliferation in their lymph nodes in response to such grafts, indicates that the thymus in some way controls the proliferation of immunoblasts. There are various possible ways in which this could be effected (see Parrott, 1967).

Details of the origin of macrophages which appear in sites of allograft rejection are unknown, but since animals which have received skin allografts display delayed-type hypersensitivity to graft antigens (Brent and Medawar, 1967), it seems reasonable to postulate that these cells would be derived from circulating recently-derived blood monocytes originating in the bone marrow, as demonstrated by Spector (1967) in other forms of delayed-type hypersensitivity. Electron microscopic studies of skin allografts (Wiener, Spiro and Russell, 1964) suggest that macrophages and other 'graft rejection cells' may destroy graft epidermis by a mechanism of close cell contact analogous to target cell destruction in monolayer cultures.

EVIDENCE FOR ALLOGRAFT IMMUNITY

The fact that second-set rejection of skin or organ allografts occurs more rapidly than first-set rejection is evidence that rejection of allografts leaves the individual in an immunized state. This form of accelerated rejection of allografts is transferable to normal animals using lymphoid cells from lymph nodes or spleen, but may or may not be transferred by means of serum from immunized animals.

The humoral antibodies which are formed after allograft rejection can be detected using a variety of techniques. Both 19S and 7S antibodies are formed in the first-set rejection, whereas the only antibody subsequently produced in most species is 7S. The 19S antibodies are more efficient in cytotoxic and haemolytic reactions involving complement fixation, whereas 7S are more potent in agglutination (Möller and Möller, 1967). Complement-fixing antibodies produced during allograft rejection are particularly sensitive to detection using the immune adherence technique. Macrophage cytophilic antibodies have also been detected in mice following skin grafting (Nelson, 1969). The humoral antibodies formed as a result of transplantation can be entirely responsible for the rejection of some types of allografts which are characterized by a high degree of sensitivity to humoral antibodies, such as cells of the lymphoid series. With non-sensitive allografts, humoral antibodies may give rise to the opposite phenomenon of immunological enhancement, whereby graft survival is prolonged. Enhanced survival of both neoplastic and non-neo-plastic cells has been recorded in antiserum-treated recipients. Whereas the survival of skin grafts is only slightly prolonged, the survival of other tissues, such as allogeneic ovary, may be greatly increased in pre-treated hosts (Möller and Möller, 1967). The degree of enhancement is not necessarily related to the titre of humoral antibody elicited by a particular tissue, and enhancement can be used as a very sensitive technique for demonstrating humoral antibodies not easily demonstrable by other techniques. Available evidence suggests that enhancement may be due to antibody coating the donor cells thereby protecting them from destruction by host immune lymphocytes; and also antibody-induced depression of the allograft reaction may occur due to antibody combining with antigenic determinants on the donor cells (Möller and Möller, 1967).

There is abundant evidence for cell-mediated immunity functioning during the rejection of allografts. *In vitro* studies indicating cell-mediated immune reactions have shown (1) the migration of peritoneal macrophages from mice or guinea-pigs sensitized by skin allografts is inhibited by culturing their peritoneal exudate cells with macrophages from the donor strain (acting as the antigen) or in the presence of semi-soluble antigen extracted from the lymphoid tissues of the homograft donor (Al-Askari, David, Lawrence and Thomas, 1965; Dumonde and Medawar, 1967); (2) sensitized lymphocytes attack and destroy allogeneic cells in monolayer culture without the apparent intervention of complement or free antibody (Möller and Möller, 1967); (3) in the mixed leucocyte reaction, the degree of lymphocyte transformation may be increased by the exchange of skin allografts between the lymphocyte donors (Bain, Lowenstein and Maclean, 1964).

Under the appropriate conditions, all the manifestations of delayed-type hypersensitivity reactions can also be reproduced in transplantation systems *in vivo*. Thus direct skin reactions having the appearance, tempo and histological characteristics of tuberculin-type reactions can be elicited by the intradermal injection of antigen in the form of living or killed donor lymphoid cells, or extracts thereof, in guinea-pigs previously sensitized by grafting with donor skin or by the footpad injection of donor thymocytes or lymphoid cells in Freund's adjuvant (Brent and Medawar, 1967). Similar reactions have been reported in humans and other species. Another type of reaction, the 'immune lymphocyte transfer reaction' (ILT), is excited in the donor animal by the intradermal injection of sensitized lymphoid cells from the recipient previously sensitized by transplantation of skin or cells from the donor. To elicit an ILT reaction, the lymphoid cells must be living and may come from regional nodes, blood or peritoneal cavity. The ILT reaction is generally similar to the direct reaction in tempo and appearance, but tends to be more indurated, better demarcated, angrier, and more prolonged than direct skin reactions elicited by soluble or semi-soluble antigenic matter (Brent and Medawar, 1967).

From the above evidence, it is apparent that allografts elicit a wide range of immunological responses including the production of humoral antibodies containing 19S and 7S, complement-fixing and non-complement-fixing, and, at least in some species, cytophilic antibodies. They also induce the development of delayed-type hypersensitivity to graft antigens.

REACTIONS TO TUMOUR ALLOGRAFTS

Tumour allografts elicit much the same spectrum of immunological responses as non-neoplastic allografts, but, in many cases, differ markedly in the manner in which they are rejected. Gorer (1956) studied the morphology of the rejection of tumour allografts, and reported that tumours which become vascularized reject in much the same way as skin allografts. In tumours which are poorly vascularized or non-vascularized, such as ascites tumours, the graft or tumour cells are invaded and destroyed by macrophages which may be the predominant or only cells present in the reaction. The reaction to some types of tumours, particularly lymphomas or leukaemic cells, are mainly of an exudative nature so that the tumour cells are killed by cytolysis before invasion and phagocytosis of the debris by macrophages takes place; and it would appear that humoral antibody alone may be adequate for the destruction of this type of graft, since lymphomatous or leukaemic grafts can be shown to fail to proliferate when transplanted to animals given immune serum alone. In the case of other types of tumours transplanted to animals given immune serum, enhanced survival of the grafted tissues may result, so that the recipient animal may be eventually killed by the growth and spread of the tumour. This form of immunological enhancement induced by passive transfer of humoral antibody is similar to that seen with non-neoplastic tissues (Möller and Möller, 1967).

In contrast to the effects of passively transferring serum, the passive transfer of cells from immunized animals may inhibit tumour growth or prevent the establishment of tumours in the recipients. Thus it has been reported that lymph node cells harvested from donors 3–5 days after grafting with tumour are effective in transferring some degree of resistance, but cells harvested more than 7 days after grafting may cause enhancement of tumour growth, although this enhancement may be inhibited by mixing the cells with small amounts of hyperimmune serum (Hutchin,

Amos and Prioleau, 1967). Similar findings of the relationship of the time between grafting and lymph node cell transfer to the effect on tumour growth have also been reported by others.

There is strongly suggestive evidence that the passive transfer of macrophages from immunized donors may be effective in inhibiting establishment or growth of some types of tumours in the recipients. Thus it has been observed by various authors that macrophage-rich peritoneal exudates are effective, but precise experiments, wherein the macrophages and lymphocytes of peritoneal exudates have been separated, have produced results of greater interest. Amos (1962) showed that peritoneal macrophages, but not peritoneal lymphocytes, were capable of transferring immunity to ascites lymphoma in mice; and Bennett (1965) found that, in the case of Sarcoma I and BP8 tumours in mice, whole populations of peritoneal cells and isolated peritoneal lymphocytes were more effective in preventing tumour take in irradiated recipients than were isolated peritoneal macrophages or lymph node cells. Moreover, prolonged culture of the macrophages *in vitro* resulted in a significant loss of their anti-tumour activity.

Changes in the phagocytic activity of the macrophage system, assessed by measuring the rate of clearance of intravenously injected carbon, during the growth of various transplanted, carcinogen-induced, viral and spontaneous tumours in mice were investigated by Old, Clarke, Benacerraf and Goldsmith (1960). These authors found that in the case of the repeatedly transplanted tumours Sarcoma 180, Carcinoma 755, Ehrlich ascites carcinoma, and Sarcoma 180 ascites, there were characteristic alterations in phagocytic activity. During the phase of early tumour development clearance rates were maximally elevated and remained increased throughout the period of rapid tumour growth. With progressive deterioration of the tumour-bearing host, phagocytic activity returned to normal or below-normal levels. A slight elevation in clearance rates was observed during the growth of spontaneous mammary tumours. Following a brief phase of depressed phagocytic activity, a striking hyperplasia of macrophages in liver and spleen occurred during the development of leukaemia induced by Friend virus. Of additional interest in these studies was the observation that animals infected with BCG exhibited retarded tumour growth and prolonged survival time. A similar protective effect of BCG infection against Sarcoma J in mice has been reported by Halpern, Biozzi and Stiffel (1963), who also observed a parallel between the response of the macrophages, in terms of phagocytic activity, and the natural resistance of various strains of mice to invasion by this tumour. These latter observations suggest that the activation of macrophages which occurs during the development of acquired anti-bacterial immunity to BCG may also endow these cells with increased ability to destroy tumour cells.

The manner in which macrophages attack and destroy tumour cells *in vivo* has not been widely studied, but Journey and Amos (1962) examined the rejection of transplanted lymphomas and sarcomas in mice using the electron microscope. Many of the lymphoma cells were phagocytosed and destroyed by macrophages, while a proportion of the lymphoma cells exhibited damage when free or in contact with macrophages. Some macrophages which had ingested lymphoma cells also exhibited signs of subcellular damage. Fewer sarcoma cells were phagocytosed by macrophages, and the macrophages did not exhibit signs of damage, but a greater proportion of the sarcoma cells appeared damaged when free or in contact with macrophages. It would appear that some of the tumour cells in these studies were susceptible to the cytotoxic effects of humoral antibodies or complement; and the finding that some of the host macrophages were sensitive to complement suggests that complement may damage a proportion of the host macrophages and the allografted tumour cells by virtue of the fact that it acts on cells carrying host antibody complexed with donor antigen (Amos, 1962).

More information on the mechanisms which may be involved in the interaction between host macrophages and allografted cells have been obtained during experiments carried out *in vitro*.

CELL REACTIONS *IN VITRO*

The reactions of macrophages with target allogeneic cells *in vitro* has not been investigated on a wide scale, and much more attention has been paid to lymphocytes as cytotoxic mediators in such systems. Interactions between macrophages from immunized mice and target cells carried out by Granger and Weiser (1964, 1966) showed that peritoneal macrophages from recipient C57 Bl/Ks mice reacted in culture with fibroblasts, tumour cells or macrophages from the donor A/J mice. The reaction involved the rapid adherence of the immune macrophages to the target A/J cells, and was followed 12–60 hours later by destruction of both macrophages and target cells. The interaction was immunologically specific; and damage, but not adherence, was inhibited by the presence of metabolic inhibitors, and was not effected by immune serum alone or by the incubation of normal macrophages with immune serum. Adherence was inhibited by trypsinization of immune macrophages, but was restored by incubating the trypsinized macrophages with immune serum. The results of these experiments indicate that there is initial adherence of immune macrophages to target cells due to the presence of specific antibody (presumably cytophilic) on the surface of the macrophage, and that this is followed by mutual damage to the macrophages and target cells which is an energy-dependent process. Hoy and Nelson (1967) studied the role of cytophilic antibody in the reaction of immune macrophages with target cells following the immunization of C57 Bl/6J mice with intraperitoneal injections of Sa 1, or with A/J skin grafts. It was found that Sa 1 cells adhered to macrophages from immunized animals, but the degree of adherence was greater when normal macrophages which had been incubated with immune serum was used. Trypsinization of macrophages from hyperimmune mice abolished their ability to take up Sa 1 cells, but this could be partly restored by hyperimmune serum. Trypsinization of normal macrophages did not inhibit their ability to take up cytophilic antibodies from hyperimmune serum, but reduced or abolished their ability to take up antibodies from serum with lower titres of cytophilic antibody. Further studies demonstrated that two classes of cytophilic antibody were present in the sera of mice immunized with sheep erythrocytes, one of which was attached to a trypsin-resistant receptor and the other to a trypsin-sensitive receptor (Nelson, 1969). These findings suggest the possibility that the effects of cell-to-cell adherence mediated by macrophage cytophilic antibody could depend on the type and amounts of cytophilic antibody involved.

Studies on the phagocytosis of ascites tumour cells by mouse macrophages *in vitro* by Bennett, Old and Boyse (1964) appeared to show that immune serum was essential for phagocytosis of tumour cells when macrophages were harvested from either immunized, normal or syngeneic mice. Concentrations of immune serum lacking cytotoxic activity still promoted phagocytosis of the target cells by macrophages. It would thus appear that some types of tumours elicit much higher titres of opsonizing antibody than other tumours, since Granger and Weiser (1964, 1966) and Hoy and Nelson (1967) did not observe phagocytosis of tumour cells in their experiments. The apparent inability of macrophages from immunized animals to endocytose target cells in the absence of immune serum reported by Bennett *et al.* (1964) may relate to the known heterogeneity of cytophilic and non-cytophilic antibodies responsible for adherence and opsonizing activity. Moreover, it is known that there are wide variations in the susceptibility of various types of tumour cells to phagocytosis by macrophages, and there are also variations in the relative cytotoxic effects of immune macrophages and immune lymphocytes in target-cell culture systems.

GRAFT-VERSUS-HOST REACTIONS

The transfer of living immunologically competent cells into an allogeneic recipient unable to reject them in the usual allograft reaction results in a syndrome referred to as the graft-versus-host (GVH) reaction. The GVH reaction can be elicited in newborn rats or mice, not yet immunologically competent, by injecting them with allogeneic lymphoid cells from adult animals; and can be elicited in adult animals whose immunological reactivity has been eliminated by irradiation, by injecting them with allogeneic adult lymphoid or bone marrow cells. After a few days, the transferred cells develop immunity against the tissue antigens of the recipient animal and react against them, thus bringing about a serious or fatal condition known as 'homologous disease' or 'runt disease'. The early symptoms are manifested by hepatomegaly and splenomegaly, and by 'runting' (growth inhibition, and loss or failure to gain weight). Later, there is atrophy of lymphatic tissues with lymphocytopaenia, dermatitis, diarrhoea, and a haemolytic anaemia. The severity of the condition depends upon the degree of histoincompatibility between donor and recipient, and upon the number of allogeneic cells transferred. The fully established runting syndrome is only seen in the milder degrees of histo-incompatibility since, in other circumstances, the condition runs an acute course leading to death of the host. In the milder forms of the syndrome, the host may recover to become a chimera containing cells of both donor and host type. It is of interest that in mice which survive for long periods after GVH reactions, there is a higher incidence of malignant tumours, particularly lymphomas and tumours composed of cells morphologically in-distinguishable from macrophages (Schwartz and Beldotti, 1965; Walford and Hildemann, 1965; André-Schwartz, Schwartz, Mirjl and Beldotti, 1967).

Studies on the histological changes in the tissues in rats dying from runt disease by Billingham, Defendi, Silvers and Steinmuller (1962) revealed that, 16 days after the injection of allogeneic cells, large numbers of pyroninophilic immunoblast-type cells appeared within the red pulp of the spleen, while the lymphocytes and lymphoblasts disappeared from the white pulp and were replaced by reticulum cells. The pyroninophilic cells later also appeared in the white pulp. In lymph nodes, lymphocytes and lymphoblasts disappeared from the cortex and were replaced by macro-phages and reticulum cells, and pyroninophilic cells proliferated within the medulla. At the same time, the cortical cells of the thymus underwent degeneration and macrophages became the predominating cells in the medulla. There would appear to be considerable variations in the nature of the cellular proliferation which occurs in the lymphoid tissues and the spleen in various species, but it would seem that the proliferation of pyroninophilic cells and reticulum cells is a fairly constant feature, and there is also the appearance of extra-medullary haemopoietic tissue in some cases. There does not seem to be definite evidence of any marked proliferation of macrophages in these tissues; this is in striking contrast to the liver where there is a marked increase in the number and size of the Kupffer cells lining the hepatic sinusoids, accompanied by similar increases in the phagocytic activity of these cells, and in the size and weight of this organ (Howard, 1961a, 1963).

Howard (1961 *a*, 1963) studied the activity of macrophages during GVH reactions between two inbred strains of F_1 hybrid mice following the injection of immunologically competent cells from either of the parental strains, and found that the increased phagocytic activity of macrophages, measured using the carbon clearance test, varied over a wide range depending on the strain combinations used. No comparable rise in phagocytic activity was observed after the injection of syngeneic hybrid spleen cells. The injection of parental strain spleen cells from a donor pre-immunized against the host (via the other parental strain) produced an intensified or accelerated reaction; whereas pre-immunization of the recipient against the donor exerted a strong inhibitory influence on the development of a rise in phagocytic activity. It was suggested that the rise in the phagocytic activity of macrophages represents a reaction of the host to destruction in lymphoid tissue attacked by donor cells.

At the height of the rise in macrophage phagocytic activity during GVH reactions, there is a very marked increase in the susceptibility of F_1 hybrid mice to the lethal effects of bacterial endotoxins (Howard, 1961*b*), a marked increase in the resistance to infection with *Diplococcus pneumoniae* (Cooper and Howard, 1961), and profound immunological depression when assessed by the formation of antibody against *Salmonella typhi* H antigen and the rejection of allogeneic skin grafts (Howard and Woodruff, 1961). The latter observations contrast markedly to the adjuvant effect of endotoxin or BCG infection wherein there is also stimulation of macrophage phagocytic activity, and this suggests that the immunological deficit manifested during GVH reactions is related to the depletion of immunologically competent host cells (Howard, 1963). It is of interest that the enhanced phagocytic activity (and mortality) during the GVH reaction is greatly reduced by splenectomy 2–8, but not 11, days after the injection of parental strain spleen cells in F_1 hybrid mice (Biozzi *et al.* 1964). These findings suggest that splenectomy removes a large focus of the GVH reaction since many of the donor cells settle in this organ, and thus also removes the source of much of the cell debris formed by the destruction of host cells.

Studies on the origin of proliferating cells during GVH reactions have shown that, whereas cellular proliferation in lymphoid tissues is largely of host origin in many combinations, when C57 Bl cells are inoculated into (C57 Bl × CBA-T6T6)F_1 recipients, the resulting proliferation is predominantly of donor origin and isolated liver macrophages yield mitoses which are largely of donor karyotype (Howard *et al.* 1965; Boak, Christie, Ford and Howard, 1968). The majority of dividing Kupffer cells in these studies were of the donor C57 Bl karyotype, irrespective of whether the donor inoculum had been spleen cells, thoracic duct cells (containing up to 10 per cent large lymphocytes), or partly-purified

thoracic duct small lymphocytes (containing 1.5 per cent large lymphocytes). The thoracic duct cells were more effective than the spleen cells.
Further studies (Howard, Christie, Boak and Kinsky, 1969) also showed
that a large proportion of mitosing alveolar and peritoneal macrophages
in GVH reactions initiated by glass wool-filtered lymph node cells or
thoracic duct lymphocytes were of donor karyotype. These studies were
limited to, and dependent upon, the examination of dividing cells only
(forming less than 2 per cent of the total cells present) and do not indicate
the proportion of donor cells in the total non-dividing populations of
macrophages. However, the absence of cells having the morphological
and functional properties of macrophages among the lymph node and
thoracic duct cells used in these studies, and the lower percentage of
donor-derived macrophages when spleen cells (containing many macrophages) were used, indicate that lymphocyte-like cells can act as the
precursor of a proportion of liver, alveolar, and peritoneal macrophages
during GVH reactions. Moreover, in syngeneic radiation chimaeras
stimulated by either partial hepatectomy or diethylstilboestrol, and in
non-irradiated histoincompatible mice kept in parabiosis and stimulated
with diethylstilboestrol, evidence has been adduced which indicates that
Kupffer cells are normally derived from the same precursor in the bone
marrow as are free monocytes and macrophages, rather than multiplying
by a process of self-replication (Howard, 1970). These latter findings again
bring to the forefront the controversial topic of the possibility of the
occurrence of lymphocyte-to-macrophage transformation (see Chapter 2).
Until more is known about the various sub-populations of lymphocytes
and lymphocyte-like cells, it would be unwise to speculate at this stage,
but it would be of great importance and interest if it could be established
that sensitized lymphocytes could transform to macrophages which
retained the sensitized state. The nature of the stimulus to macrophage
proliferation during GVH reactions, irrespective of the cells being of host
or donor origin, is unknown; moreover, the length of survival and fate
of the macrophages of donor origin in those animals which survive the
GVH reaction has not been established.

It is not known if the proliferating macrophages play some part in
damaging the cells of the host during GVH reactions, but there is evidence
that macrophages appear to be able to initiate allogeneic reactions in
some cases. Weiser *et al.* (1965) immunized C57 Bl/Ks mice with Sa 1
tumour from *A* strain mice, and found that the injection of immune
peritoneal cells from the C57 Bl/Ks mice caused the death of a large
proportion of *A* strain recipients within 3 days. This form of 'acute
allogeneic disease' was also produced by injecting peritoneal macrophages
alone, but not by injecting peritoneal lymphocytes alone or cell-free
peritoneal fluid. Carbon-labelled donor macrophages were seen within

the pancreas, and death appeared to be due to an acute pancreatitis. These findings strongly suggest that the immunological specificity observed in these experiments can only be attributable to the presence of cytophilic antibody on the surface of the donor macrophages, and relates to the *in vitro* findings on macrophage–target-cell interactions by Granger and Weiser (1964, 1966) cited previously. The pathological features of pancreatitis in 'acute allogeneic disease' induced by immune macrophages differ so markedly from the histological features of GVH reactions produced by the transfer of immunologically competent lymphoid cells that it is not possible to suggest that the former is evidence that macrophages can initiate true GVH reactions. Moreover, there is little doubt that, in mammals, GVH reactions are mediated by donor small lymphocytes which rapidly transform to large pyroninophilic cells within the host tissues (Gowans, 1962). It has been reported that runt disease may be produced in irradiated F_1 hybrid hosts by the injection of oil-induced peritoneal exudates (Cole and Garver, 1961); however, although such exudates contain a predominance of macrophages, it is well-recognized that small lymphocytes are also constantly present and cannot be excluded as the mediators of any reaction.

CELLULAR RESISTANCE TO SYNGENEIC AND AUTOCHTHONOUS TUMOURS

EVIDENCE FOR SPECIFIC TUMOUR IMMUNITY

There is evidence in animals and man that there are tumour-specific antigens of the transplantation type, that there is a reaction of the host against spontaneous, induced, or transplanted syngeneic tumours, and that this reaction is in some cases mediated by cells in a manner similar to delayed-type hypersensitivity reactions and allograft rejection.

The earliest studies indicating the presence of specific tumour antigens were made by Foley (1953) and repeated by Prehn and Main (1957): these authors showed that primary tumours, induced in mice of a pure strain by the chemical carcinogen methylcholanthrene, are immunogenic when transplanted into inbred mice of the same strain and behave as if they had the properties of an allograft, in that they possess weak antigens which are absent in the host animals. These findings were later confirmed by others using different carcinogens in rats and mice. In pure strains of animals, syngeneic tumours are readily transplantable to animals of the same strain, but unlike syngeneic grafts of skin or other tissues they give rise to a state of immunity even in syngeneic recipients. The production of a state of immunity may be shown to exist in syngeneic recipient animals immunized with (1) tumour grafts subsequently removed surgically, (2) living tumour cells rendered incapable of indefinite growth by exposure to irradiation or cytotoxic chemicals, or (3) tumour cells in numbers too small to give rise to subsequent tumour growth. Doses of tumour cells which would normally produce tumours in non-immunized animals fail to produce tumours in animals immunized by one of these techniques. These techniques are limited to pure-line strains of rats and mice, but they have demonstrated that tumour-specific antigens are present in many different types of tumours (carcinomata, sarcomata, lymphomata, etc.), and the carcinogenic agents used for tumour induction include most of the well-known chemical and physical carcinogens, oncogenic viruses, irradiation, and physical agents such as inert plastic films (Alexander and Fairley, 1967). The tumour-specific antigens are weak, and the immunity elicited is usually only sufficient to cause the rejection of a small challenging inoculum of tumour cells. Moreover, each type of tumour generally possesses its own specific antigen, and there is no cross-immunity affording protection to a different type of tumour other than the one used for immunization.

As in the case of allograft immunity, tumour immunity can be transferred by the injection of lymphocytes or macrophages from animals immunized against the tumour. In this connection, it is of interest that it has been found that peritoneal cells (predominantly macrophages) are more effective than lymph node cells in producing suppression of the growth of some types of tumours (Old, Boyse, Bennett and Lilly, 1963). Systemic passive transfer of resistance to tumour growth has been achieved with lymph node cells, although this has not been consistently successful (Old, Boyse, Clarke and Carswell, 1962), but suppression of tumour growth can be achieved by the injection of lymph node or spleen cells from immunized donors together with tumour cells in syngeneic recipients (Rosenau and Morton, 1966; Mikulska, Smith and Alexander, 1966). Moreover, tumour cells which have been mixed with immune lymphocytes (from a donor immunized with the tumour) lose their capacity to grow on being injected into syngeneic recipients (Klein, Sjögren, Klein and Hellström, 1960), while non-immune lymphocytes fail to inhibit tumour growth under similar circumstances. It is doubtful if, even when present, circulating antibodies cytotoxic to tumour cells play a significant part in the resistance of immunized animals against the solid tumours, since it has been generally shown that immune serum of syngeneic origin may produce enhancement of tumour growth (Old *et al.* 1963; Möller and Möller, 1967). However, in the case of transplantable syngeneic leukaemias, antisera produced in allogeneic strains of mice have been shown to provide excellent protection against subsequent tumour inoculation, but they produce very limited protection as therapy administered after tumour inoculation (Gorer and Amos, 1956).

All the available data are consistent with the view that a direct cytotoxic action of immune lymphocytes is mainly responsible for the resistance of animals against syngeneic tumour-specific antigens (Alexander and Fairley, 1967). Lymphocytes are also effective with autochthonous primary tumours, and it has been shown that the growth of chemically induced fibrosarcomata in the rat can be retarded by the injection of thoracic duct lymphocytes from normal syngeneic or allogeneic rats immunized with biopsy material from the tumour to be treated (Delorme and Alexander, 1964). The tumour-inhibiting action of the immune lymphocytes is limited to the immunizing tumour alone, which suggests that the lymphocytes are directed against specific antigens which are different for each tumour. Although the immune lymphocytes are cytotoxic against the fibrosarcoma cells *in vitro*, additional findings suggest that their action *in vivo* is indirect and that they cause the tumour-bearing host to react by stimulation with a subcellular factor, possibly RNA (Alexander, Delorme and Hall, 1966; Alexander and Fairley, 1967).

NON-SPECIFIC TUMOUR IMMUNITY

It has often been suggested that tumours arise because the immune mechanism of the host is impaired. Although this may be a contributory factor, the fact that only very small numbers of tumour cells may be necessary for the successful transfer and growth of antigenic tumours in normal syngeneic recipients suggests that impaired immunity is not essential for primary autochthonous tumours to arise *de novo*. However, there is evidence that the immune mechanism may become altered during the growth of established primary tumours. For example, there is experimental evidence that an actively growing primary tumour exhausts the supply of host anti-tumour lymphocytes, but that, if the tumour is surgically removed, the concentration of cytotoxic lymphocytes builds up and the animal can then reject a subsequent tumour autograft (Mikulska *et al.* 1966). Moreover, there is abundant evidence that, in patients with malignant lymphomata, myeloproliferative diseases, carcinomas and sarcomas, there is often impaired delayed-type hypersensitivity to antigens encountered before the onset of the disease (such as tuberculosis, mumps, etc.), in addition to a reduced capacity to develop a delayed-type hypersensitivity reaction to a new antigen (see Alexander and Fairley, 1967). Patients with malignant disease sometimes exhibit temporary or indefinite delay in the rejection of allografts, and it has been reported that growth and subsequent metastasis of tumour allografts can occur in patients with advanced cancer (Southam, Moore and Rhoads, 1957). Tumour-bearing animals also exhibit diminished immune responses, and a reduced capacity to reject weakly histoincompatible skin allografts (Prehn and Main, 1957).

RELATION OF MACROPHAGE ACTIVITY TO TUMOUR GROWTH

A number of studies have shown that the phagocytic activity of the macrophage system may become altered during the progression of malignant disease in animals and man. Sheagren, Block and Wolff (1967) assessed the rate of clearance of intravenously injected radio-iodinated heat-aggregated human serum albumin in patients with Hodgkin's disease, and found that, in comparison with control subjects, clearance was accelerated in patients with advanced disease. Salky, DiLuzio, Levin and Goldsmith (1967) assessed the rate of clearance of intravenously injected lipid emulsions, and also reported an increased rate of clearance in patients with malignant diseases. In contrast to these findings, Quantock (1969), also using radio-iodinated heat-aggregated human serum albumin, found that phagocytic activity was apparently reduced significantly in 76 patients with malignant disease when compared to 58 controls of the same age group; however, further analysis of the results showed that there was no significant difference between the clearance rates of the control group and the patients who had malignant disease with no evidence of metastases, but there was a marked reduction in phagocytic activity in patients who had clinical evidence of metastases which accounted for the overall differences between control and malignant groups.

Old *et al.* (1960) studied phagocytic activity, using the carbon clearance technique, in mice with spontaneous, methylcholanthrene-induced, virus-induced, and syngeneic transplant tumours. In the case of Friend virus leukaemia, there was a marked and progressive rise in clearance rates and a rise in the weight of the liver and spleen as the condition developed, but there was a terminal phase of progressive fall in clearance as the leukaemia spread prior to the death of the animals. Slightly increased clearance rates were observed in mice during the growth of spontaneous mammary adenocarcinomata, and slight increases were also observed with syngeneic tumour implants and methylcholanthrene-induced sarcomata. It was also found that mice infected with BCG or treated with zymosan (a known stimulant of macrophage phagocytic activity) were protected against challenge with Sarcoma 180; and survival time was increased two-fold in BCG-infected hosts inoculated with Ehrlich ascites, and also prolonged in the case of Carcinoma 755. A similar protective effect of BCG infection against Sarcoma J in mice was observed by Halpern, Biozzi and Stiffel (1963); and these authors also noted that strain differences in susceptibility to invasion by this tumour were related to macrophage phagocytic activity, in that (1) marked stimulation of macrophage phagocytic activity accompanied the growth of grafted Sarcoma J in C57 Bl strain in which this tumour spontaneously regresses in a high percentage of cases, (2) phagocytic activity was not altered in C3H and 'Swiss' strains which are less resistant and eventually all die from the effects of tumour growth, and (3) a significant reduction of phagocytic activity occurred in F_1(C3H × C57 Bl) hybrids which exhibit the least resistance to invasion by Sarcoma J. These findings may be related to those of Stern (1960) who found that the uptake of colloidal gold ^{198}Au by the macrophages of the liver and spleen was lowest in those strains which exhibited the highest incidence of spontaneous tumours, and highest in those with the lowest incidence of spontaneous tumours. Stern postulated that the initial step in carcinogenesis may be a 'weakness' of macrophages, either genetically determined or resulting from endogenous or environmental insults, which interferes with their ability to handle macromolecules concerned with the regulation of normal growth: the degree of macrophage 'damage' could also be a factor in determining the rate of growth and spread of the neoplastic disease. While these hypotheses have remained speculative, they nevertheless offer an attractive basis for discussion and future research in this field. It is possible that the better survival associated with the increased phagocytic activity of macrophages may not be entirely due to a direct anti-tumour cell effect of the host macrophages, but may be due to the macrophages taking up and 'de-toxifying' debris and 'toxins' liberated during necrosis of tumour cells.

RESISTANCE TO METASTATIC SPREAD

The occurrence of 'sinus histiocytosis' (the extensive proliferation of macrophages in the sinuses of lymph nodes) in the lymph nodes draining the sites of tumours, has been considered by several authors as being a favourable prognostic sign in a variety of tumours in animals and man. Carter and Gershon (1966, 1967) have shown that in hamsters grafted with the transplantable lymphoma ML which metastasizes widely, the lymph nodes draining the graft did not exhibit sinus histiocytosis, and ML cells entering the nodes multiplied and were not phagocytosed. In contrast, grafts of the similar lymphoma NML rarely metastasized, the nodes draining the grafts exhibited marked sinus histiocytosis and NML cells reaching the nodes were phagocytosed by macrophages. If the tumours were grafted together, sinus histiocytosis occurred, but only NML cells were phagocytosed. The presence of macrophages in large numbers did not appear to affect the growth of ML cells, but a lack of lymphoid cells in the nodes resulted in reduced phagocytosis of NML cells.

A number of studies have been made in man on the relationship of sinus histiocytosis in the lymph nodes draining carcinomata of the breast, stomach and colon to tumour spread and ultimate survival. Many authors have concluded that sinus histiocytosis signifies a better prognosis, especially if it is associated with mononuclear infiltration of the primary tumour (Black and Speer, 1958, 1960; Cutler *et al.* 1966; Wartman, 1959). In contrast, Berg (1965) found that the occurrence of sinus histiocytosis was of no value in assessing prognosis in breast cancer. These observations are still the subject of controversy. Sinus histiocytosis in lymph nodes occurs in the absence of malignant tumours, and reactive changes of some form are almost always present in the lymph nodes draining the site of any malignant tumour, being particularly marked in those cases where necrosis in the tumour is widespread. It would be difficult to persuade most histopathologists that prognostic significance can be ascribed to such frequent changes, particularly if they are present when the tumour may have already have metastasized widely. The significance of infiltration of malignant tumours by lymphoid cells probably merits further attention in the human, since it has been shown that the presence of lymphoid infiltration in carcinoma of the testis indicates a more favourable prognosis. In this connection, it would appear that, in the case of many solid tumours in experimental animals, anti-tumour immunity is mediated via lymphocytes; whereas macrophages appear to be more concerned in defence against leukaemias and lymphomata.

AUTO-IMMUNITY

It is now recognized that immune reactions against one's own tissues is a frequent occurrence, and this phenomenon may be observed in association with various pathological disorders, also in circumstances where it may have some pathogenic significance, and in healthy persons where it may have no pathological significance whatsoever. In those circumstances where auto-immunity may play a part in the pathogenesis of various clinical or experimental diseases, these diseases may be divided into three classes: they are (1) those in which auto-antibodies are directed against circulating blood cells; (2) those in which the auto-antibodies are principally directed against a specific tissue or organ; and (3) a heterogeneous group of disorders in which various immunological abnormalities, including the presence of auto-antibodies, may be demonstrable, and these are often grouped together as the 'collagen diseases' or 'immune disorders of connective tissues'.

AUTO-ANTIBODIES AGAINST CIRCULATING BLOOD CELLS

The role of macrophages in the removal of aged or effete erythrocytes was presented in Chapter 5, together with evidence that in auto-immune haemolytic anaemias, and in neutropaenias and thrombocytopaenias due to auto-antibodies, it is the macrophages of the liver and spleen which are primarily responsible for the removal of the cells damaged by auto-antibodies directed against them.

AUTO-ANTIBODIES DIRECTED AGAINST SPECIFIC ORGANS OR TISSUES

Among the diseases in man which have been shown to have a definite auto-immune basis are Hashimoto's thyroiditis, pernicious anaemia, ulcerative colitis, primary biliary cirrhosis, active chronic (lupoid) hepatitis, cryptogenic cirrhosis and ideopathic Addison's disease. Antibodies reacting with tissue antigens have also been described in a variety of less well-documented conditions, where humoral antibodies may also play an important or predominant role in their pathogenesis. Among the latter are sympathetic ophthalmia and other eye disorders, skin disorders such as psoriasis and the effects of burns, multiple sclerosis, myasthenia gravis, primary atypical pneumonias and emphysema, chronic pancreatitis, rheumatic carditis and acute glomerulonephritis. Auto-antibodies may also be demonstrable after damage to tissues, such as with burns, infections such as tuberculosis, or infarction of tissues.

The role of macrophages in the pathogenesis of these conditions is largely unknown and uninvestigated. While these cells are constantly present among the cellular infiltrate characterizing the lesions in such conditions as Hashimoto's thyroiditis or auto-immune liver diseases in man, it is not known whether they are present as a primary pathogenic feature or in response to tissue damage by sensitized lymphoid cells or humoral antibodies. Macrophages constantly form a proportion of the early cellular infiltrate in experimental auto-immune diseases, such as experimental allergic encephalitis and thyroiditis (Kosunen, Waksman and Samuelsson, 1963; Kosunen and Flax, 1966). The damage to thyroid tissues in early experimental thyroiditis has been reported as being due to the invasion of thyroid tissues and penetration of follicular cells by macrophages and lymphocytes, but in the later stages of the disease process, lymphocytes predominate and blast cells and plasma cells are also numerous (Roitt, Jones and Doniach, 1961).

There is much evidence that delayed-type hypersensitivity reactions are operating in many of the auto-immune diseases induced in experimental animals (see Roitt and Doniach, 1967). The possible role of macrophages in delayed-type hypersensitivity has been considered earlier in this chapter. In human auto-immune disorders, the evidence for a state of delayed-type hypersensitivity is less impressive, and in many instances the data are consistent with an active role of circulating antibodies and sometimes of immune complexes in the pathogenesis of these diseases.

THE COLLAGEN DISEASES

The diseases which are commonly grouped under this heading include rheumatoid arthritis, systemic lupus erythematosus (SLE), polyarteritis nodosa, scleroderma and dermatomyositis. Other, rarer conditions, such as Sjögren's syndrome, may also be included in this group. This heterogeneous group of diseases have the following features in common: (1) there is a widespread degenerations of collagen, with 'fibrinoid necrosis' which is most marked in SLE and polyarteritis nodosa, and least conspicuous in scleroderma; (2) the lesions affect many systems in the body, and there is often a profound constitutional effect; (3) immunological abnormalities are usually demonstrable; (4) there is an interrelationship between the disorders, in that some cases of rheumatoid arthritis develop features of SLE, a polyarteritis may complicate SLE or rheumatoid arthritis, an inflammatory arthropathy may be present in any of these disorders, and so on; and (5) the aetiology and pathogenesis of these conditions is still largely unknown.

The immunological nature of these diseases is indicated by the presence of circulating antibodies directed against various autochthonous antigens. Thus, in rheumatoid arthritis, the presence of 'rheumatoid factor' (RF) may be demonstrated by a variety of techniques. RF is an immunoglobulin reacting with 7S globulin (IgG), and RF itself usually belongs to the 19S (IgM) fraction, but there are similar factors belonging to the IgG and IgA groups of immunoglobulins. A variety of other factors may also be demonstrable in the serum in rheumatoid arthritis; whereas most are autospecific and directed against the patient's own gammaglobulins, isospecific and heterospecific factors which react against determinants not present in the patient's own gammaglobulins have also been described. The latter serve to illustrate the diverse immunological abnormalities which may be present. In SLE there is characteristically present a number of antibodies directed against a variety of autochthonous nuclear antigens, but it is also common to find antibodies against leucocytes, platelets and colon antigens, and rheumatoid factor is also often present. In scleroderma, dermatomyositis and Sjögren's syndrome, a similar range of autoantibodies as in SLE is often present, and thyroid antibodies are also demonstrable with some frequency.

In the inflammatory arthropathy which characterizes rheumatoid arthritis, and which may also be present in the other collagen diseases, there is hyperplasia of the surface synovial cells, including the macrophages (Type A cells) and multinucleate giant cell forms are frequently seen in this layer. In the sub-synovial tissues, lymphocytes and plasma cells are present, often arranged in nodular and perivascular fashion, but macrophages and polymorphs may also be present in variable numbers. The proliferating rheumatoid pannus, which spreads across the joint surface and replaces the articular cartilage, contains large numbers of macrophages. The histological appearances of the joint tissues are occasionally similar to those seen in delayed-type hypersensitivity reactions in the skin.

There is strong evidence that antibody is produced locally by plasma cells in the synovial tissues (Norton and Ziff, 1966), but the appearances are not those of a local allergic reaction mediated by antibody. In actively inflamed joints, polymorphs predominate in the synovial fluid, and may contain ingested complexes of RF and IgG (Hollander, McCarty, Astorga and Castro-Murillo, 1965). The injection of purified IgG from a rheumatoid patient into a non-inflamed joint in another rheumatoid patient having circulating RF results in acute inflammation in the injected joint (Hollander *et al.* 1966). This suggests that the acute changes in the joints in this disease may have an Arthus-type component, whereas the chronic arthritis may involve some form of cell-mediated immune reaction, possibly involving delayed-type hypersensitivity.

7

THE ROLE OF MACROPHAGES IN IMMUNOLOGICAL RESPONSES

INTRODUCTION

Immunity in its broadest sense is divisible into three stages. (1) There is initial *recognition* of antigenic material which is foreign, not-self or otherwise immunologically unacceptable. (2) This is followed by the induction of an immune *response* which culminates in the production of humoral and/or cell-bound antibodies. (3) There may be the subsequent occurrence of an immunological *reaction* between antibodies and the antigenic material. There is now abundant evidence that macrophages are not only involved in immunological reactions of various kinds (see Chapter 6), but play a major role in the recognition of foreign material (see Chapter 3) and in the subsequent induction of the immune response.

The concept that phagocytic cells with some of the morphological characteristics of macrophages are capable of synthesizing antibodies has received little general support. Schaffner and Popper (1962) found that in active post-necrotic cirrhosis of the liver there are cells having phagocytosed material in one part of their cytoplasm, and, generally separated by the nucleus, in the other part abundant rough endoplasmic reticulum in which gamma globulin could be demonstrated immunocytochemically. They suggested that gamma globulin and antibody formation is stimulated by, and directed against, material incorporated by phagocytosis. Similar cells, each with one half containing many lysosomes and the other half containing abundant rough endoplasmic reticulum, have also been described as being present in the granulomatous reaction in the draining lymph nodes and foot-pads of guinea-pigs injected with Freund's complete adjuvant into the latter site (Pernis, Bairati and Milanesi, 1966). The significance of these findings is not clear since the cells described certainly do not have the typical morphology of macrophages. Other conclusions that macrophages are capable of synthesizing antibodies have been largely based upon experiments carried out using peritoneal fluid exudates in which, although the macrophage was the predominant cell, variable numbers of lymphocyte-like cells were also present. Subsequent studies have established that antibodies are synthesized by the lymphocytes, and not by the macrophages, in such cell populations. The presence of large amounts of antibody within or attached to the surface of macrophages has also been interpreted as evidence of antibody synthesis by these cells, but it is now established that macrophages may contain antibody which is synthesized by lymphoid cells.

Other difficulties have related to determining the exact identity of various cell types among mixed populations of cells. The introduction in recent years of a variety of elegant techniques has enabled the identification of individual antibody-producing cells even when large numbers of other cell types may be present. Thus it has been shown that in the early phase of immune responses, antibodies are produced by a variety of cells, namely lymphocytes, large pyroninophilic cells and actively proliferating 'blast' cells, which have similar rather primitive ultrastructural characteristics in that they possess numerous ribosomes, often clustered as polyribosomes, but little rough endoplasmic reticulum. It would appear that all these cells are probably in the process of differentiating towards plasma cells, but possess the capacity to synthesize specific antibody before they develop the characteristic morphology of the mature plasma cell. The latter possesses abundant rough endoplasmic reticulum which is a feature of cells producing protein for export. Present evidence clearly establishes the plasma cell as being predominantly responsible for the production of antibody in both primary and secondary immune responses, and definitely excludes the macrophage from being a cell capable of antibody synthesis. However, there is increasing evidence that macrophages play an important intermediate role in the events which finally lead to antibody production.

It is widely recognized that macrophages avidly ingest both particulate and soluble antigenic material which has been introduced into the body by various routes, and much evidence has been produced that immunologically-competent cells (cells capable of synthesizing antibody) are directly or indirectly stimulated to antibody production by a product, possibly RNA or RNA complexed with antigen or fragments of antigen, of the 'processing' activity of macrophages. However, there is increasing evidence which now suggests that most of the antigen taken up by macrophages is digested and is probably inactive in immune induction, but that a small proportion of antigen is carried in unprocessed form on the cell membrane of macrophages in such a way that it is 'exposed' to lymphoid cells in a manner which is much more effective than can be achieved by free antigen alone.

The production of specific antibody under normal conditions appears to be the special concern of cells situated in the lymph nodes and spleen, with other lympho-reticular tissues such as the thymus and bone marrow playing a less direct role. These tissues are, under normal conditions, predominantly composed of lymphocytes, macrophages and plasma cells; and one can assume that the arrangement of the cells and the general construction of the organs are such that the activity of the cells can be co-ordinated to maximum effect in carrying out the various functions of the organs and cells. Further morphological evidence that these particular cells play some form of co-operative role in antibody production is suggested during the production of antibody in tissues not normally directly concerned in the immunological response to antigen, such as the skin or thyroid gland, which is always preceded by the focal accumulation of lymphocytes, macrophages and plasma cells. Evidence that the lympho-reticular tissues have some form of compartmental arrangement in relation to their immunological functions is afforded by the lymph nodes, where the cells of certain regions have been clearly shown to be thymus-dependent (Parrott, 1967). It is evident that an examination of the morphological changes which occur in the spleen and lymph nodes during the primary and secondary immune responses can provide some information on the individual roles of the cells involved.

MORPHOLOGICAL CHANGES ASSOCIATED WITH ANTIBODY PRODUCTION

THE PRIMARY RESPONSE

The cellular changes in the lymph nodes of rabbits after the injection of typhoid antigen into the foot-pad was examined by Ehrich, Drabkin and Forman (1949). Twenty-four hours after injection, the sinuses contained many macrophages and polymorphs, and there were many actively proliferating pyroninophilic cells in the medulla and in the cortex adjacent to the sinuses. By the second day, swollen vacuolated macrophages were the predominant cell present in the sinusoids, and actively dividing pyronino-philic cells were very prominent in the medulla and cortex. By the fourth day, the reaction in the sinuses had largely subsided, but in the medullary cords and cortical tissue adjacent to the sinuses mature plasma cells predominated in many places although actively proliferating lymphoid cells were still present. Early germinal centres were also apparent. By the fifth day, the cortex and medulla were crowded with plasma cells together with some proliferating pyroninophilic lymphoid cells, and germinal centres were then prominent and exhibited proliferating lymphoblasts and medium lymphocytes. The efferent lymph of the node contained mainly small lymphocytes, together with some pyroninophilic cells and plasma cells. Thereafter, the numbers of plasma cells in cortex and medulla declined, until few were evident by the ninth day. On the other hand, the germinal centres had become very prominent with many dividing lymphoblasts and medium lymphocytes, and contained many macrophages which had ingested nuclear debris. The cells of the efferent lymph at this stage comprised small lymphocytes only. Maximum antibody titres in the nodes were found between the fourth and sixth days after antigen administration, and corresponded with the presence of maximum numbers of plasma cells, and also with the peak of increased RNA content of the nodes.

The changes taking place in the spleen, lymph nodes, liver, lungs, kidneys and bone marrow of rabbits after the intravenous injection of various particulate and soluble antigens was examined by Marshall and White (1950). The reactions were of a uniform type for all the antigens tested. The changes observed were largely confined to the spleen. Within 8 to 12 hours of antigen injection there was a diffuse infiltration of the red pulp with polymorphonuclear leucocytes, accompanied by the disappearance of small lymphocytes from the same area. These changes persisted for 2 to 3 days. At about 3 to 4 days, small foci of actively proliferating large pyroninophilic 'activated reticulum cells' appeared diffusely throughout the red pulp; and at a later stage the majority of cells comprising these foci consisted of 'plasmablasts' (probable plasma cell precursors). At this stage, mitotic activity was apparent in the centres of Malpighian bodies, and in the course of the following 3 to 4 days numerous stellate large cells with basophilic cytoplasm appeared in the area and were intermingled with large lymphocyte-like cells. The proliferating tissue caused the appearance of a pale area, the so-called germinal centre. The cytology of the germinal centres closely resembled that of Flemming's centres seen in human material in reactive follicular hyperplasia, but was not identical, since macrophages containing stainable material ('tingible body macrophages') were present in small numbers only and the characteristic 'starry sky' appearance seen in human material was absent. Germinal centre formation was confined to the spleen, and there was no comparable change in the lymph nodes examined. No changes occurred in the other organs examined. Antibody synthesis was detected within 3 or 4 days of antigen injection, at the time when cell proliferation had not progressed much beyond the stage of the 'activated reticulum cell' or 'plasmablast', and before increased numbers of mature plasma cells had become evident. Silver impregnation studies did not reveal any evidence that proliferation of metalophil macrophages of the spleen took place during primary immunization, but alteration in the morphology of the cells was observed. Thus the number of finely branched macrophages decreased and coarsely branched amoeboid type cells took their place. There was also an increase in the number of free macrophages within splenic sinusoids. In the marginal zone at the junction of white pulp (Malpighian bodies) and red pulp, a true increase in the number of macrophages appeared to take place.

In regional lymph nodes after the subcutaneous injection of antigen, there is a true proliferation of metalophil macrophages of the sinuses, and this closely resembles the 'sinus catarrh' seen in inflamed human lymph nodes (Marshall, 1956). The failure of a similar proliferation in the sinusoidal macrophages of the spleen indicates that the lymph node reaction is non-specific and is probably related to the absorption of products of dead or damaged tissues at the site of injection rather than to the action of antigen *per se*.

From the above and subsequent studies, it would appear that the morphological changes which occur during primary immunization are largely confined to lymph nodes or spleen depending respectively upon whether the subcutaneous or intravenous route of administration of antigen is utilized. The principal changes may be summarized as (1) an initial short-lived polymorphonuclear and macrophage reaction in the sinuses of lymph nodes or venous sinuses of the spleen; (2) a rapid proliferation of primitive-type basophilic pyroninophilic cells ('immunoblasts') in the medulla and superficial cortical areas of lymph nodes or in the red pulp of the spleen; (3) the appearance of immature and mature plasma cells by the fourth day; and (4) germinal centre formation in lymph nodes or spleen beginning about the fourth day and reaching maximum size as plasma cells elsewhere in the organs are diminishing in number and antibody production.

Special techniques for identifying individual antibody-producing cells (Jerne, Nordin and Henry, 1963) demonstrate a marked increase in such cells within one or two days after antigen administration, and they reach maximum numbers four or five days after immunization. The findings also confirm that the earliest antibody-producing cells have a varied morphology, and that plasma cells are productive at a later stage. The fact that levels of antibody in the serum estimated by the usual methods do not exhibit a detectable level until three or four days after immunization has led to the concept of a 'lag phase' during which the *induction* of immunity, to be contrasted with the *production* of immunity associated with the appearance of antibodies, takes place. The advent of more sensitive techniques for detecting serum antibodies and antibody-producing cells has revealed that antibody levels may begin to rise as early as 24 hours after primary immunization, indicating a much shorter inductive phase than was originally thought to exist.

THE SECONDARY RESPONSE

Marshall and White (1950) observed that, if repeated intravenous injections of various antigens are given at three- or four-day intervals, the foci of plasmablasts which they had observed appearing in the red pulp of the spleen during primary immunization increased rapidly in size, mitoses being frequent, so that they occupied the greater part of the splenic pulp in some cases. After three or four weeks of repeated injections, the foci were composed of mature plasma cells alone, but immature plasma cells were still present in the marginal zone. The germinal centres of the Malpighian bodies exhibited progressive diminution in activity, until after three or four weeks it became difficult to distinguish them from the surrounding small lymphocytes. Fagraeus (1948) also noted a marked increase in the number of plasma cells in the red pulp of the spleen after the administration of intravenous antigen, but did not record any change in the Malpighian bodies. She also noted the changes in lymph nodes after the subcutaneous administration of antigen, and observed a marked increase in the number of immature and mature plasma cells in the medulla with an associated increase in the cortex adjacent to the sinuses.

Both Fagraeus (1948) and Marshall and White (1950) reported the presence of focal accumulations of plasma cells, and sometimes of immunoblasts, in other tissues after repeated antigen administration. These were observed in the lungs, liver, bone marrow and renal pelvis.

The morphological changes which occur during the secondary response may be summarized as (1) a prolonged proliferation of plasma cell precursors resulting in an increase in the number of mature plasma cells in lymph nodes and spleen; (2) an eventual progressive diminution in activity of germinal centres in lymph nodes and spleen when no more antigen is available; and (3) focal accumulations of lymphoid cells in other organs.

CHANGES IN LYMPH-BORNE CELLS DURING THE IMMUNE RESPONSE

Hall and his co-workers have carried out extensive studies on the changes in the cell populations of afferent and efferent lymph of popliteal nodes of sheep given subcutaneous injections of antigens into the lower leg (Hall and Morris, 1963; Hall, 1967). Normally, the *afferent* lymph contained 100–3,000 white cells per mm³, and about 2 million cells passed into the node every hour by this route. About 80 per cent of the cells were mature lymphocytes, 10 per cent were macrophages and monocytes, and 10 per cent were neutrophil polymorphs. Within a few hours of antigen injection, the number of neutrophil polymorphs increased until they comprised half of the cell population, but returned to pre-injection levels 20–30 hours later. Large lymphocytes and transitional cells appeared; many were highly basophilic and mitoses were common. Plasmablasts were also present but mature plasma cells were rarely seen. Macrophages and monocytes remained constant throughout the response. Normally, the *efferent* lymph contained 5,000–20,000 cells per mm³ and nearly all the cells were mature lymphocytes, and 10–75 million cells left the node by this route every hour. After antigen injection, there was a sudden transitory depression in cell output. Cell output then increased, and 24 hours after injection had doubled or trebled due to an increase in mature small lymphocytes. Later, large numbers of large lymphocytes and transitional cells appeared, followed by cells of the plasma cell series. The output of cells exceeded 300 millions per hour at the height of the response in some cases. The response to a second dose of antigen occurred more rapidly, was more vigorous, and was always shorter than the first.

It seems probable that in most mammals, lymphocytes pass in the blood stream into lymph nodes where they constantly circulate from blood to lymph by traversing specialized vascular endothelium in the cortices of the nodes (Gowans, 1966). Under normal conditions, over 95 per cent of lymphocytes in efferent lymph are derived in this way, and only about 5 per cent are actually produced in the nodes. Hall and his colleagues have hypothesized that, by this mechanism, a lymph node is provided with a large and heterogeneous population of lymphocytes from which those few with appropriate immunological potentials may be selected and retained in the node to react with whatever antigenic material has been localized in phagocytic cells (Hall, 1967). They have concluded that the early increase in the output of mature lymphocytes in the efferent lymph after immunization can only be accounted for by an increased rate of lymphocyte recirculation, and postulate that this period could be regarded as a phase of 'recruitment' of appropriate cells in terms of a clonal selection hypothesis. They consider that the large numbers of 'immunoblasts' in the efferent lymph at a later stage are probably concerned with propagating the immune response throughout the body, but the final fate of these cells is obscure.

The macrophages of the afferent lymph appear to be completely filtered out of the afferent lymph within the lymph node but their final fate is unknown. If macrophages play a role in effectively transporting soluble or insoluble antigen to lymphoid cells, and there is evidence that this mechanism is operative, this would help to account for the complete filtration of the cells from the lymph within lymph nodes. The absence of an increase in the numbers of macrophages in the afferent lymph after antigen injection indicates that only a small proportion of the many macrophages which usually appear at the local site of injection of antigen migrate to the draining node during the time of the immune response, and suggests that the majority remain sequestered at the injection site to ingest and digest the foreign material (antigen) in that area. Evidence will be quoted later that a small proportion (less than 10 per cent) of unprocessed antigen is utilized in immune induction mediated by macrophages, whereas over 90 per cent of ingested antigen is rapidly digested by the cells in the phagosome–lysosome system.

THE DISTRIBUTION OF INJECTED ANTIGENS

CELLULAR DISTRIBUTION

Early studies on the fate of injected antigens relied on the ability to recognize the characteristic shape of the antigenic particle or attached marker. Thus investigations on the fate of various cocci showed that some became localized within macrophages, but many were also initially taken up by polymorphonuclear leucocytes. Similar studies were carried out using virus particles, the tissues being examined with the aid of the electron microscope. The latter instrument was also used to locate antigen 'labelled' with a ferritin marker. In other studies, various coloured dyes coupled to antigens were traced using histological techniques. It may be concluded from these studies, and from knowledge obtained in subsequent studies using techniques which have enabled the more accurate and sensitive location of antigens, that the above methods give an inaccurate and incomplete picture of the distribution of antigen. However, the advent of fluorescent antibody techniques and of isotopically-labelled antigens has enabled the accurate location of very small amounts of antigen at both cellular and subcellular levels.

Using immunofluorescence, Coons and his associates traced pneumococcal capsular polysaccharides after intravenous injection in the mouse, and located the antigen within macrophages of the red and white pulp of the spleen; in Kupffer cells and, to a lesser extent, the parenchymal cells of the liver; in the subcapsular and medullary sinuses of lymph nodes; and in macrophages in the submucosa of the alimentary tract and connective tissue stroma of cardiac and skeletal muscle. The antigen was also located within steroid-producing cells in the adrenal cortex, ovary and testis. The fact that pneumococcal polysaccharide did not localize exclusively to macrophages in this study may relate to the very rapid diffusibility of this material, and also may in some way relate to the immunological paralysis produced by the polysaccharide whereby high doses not only fail to immunize but also inhibit the immunizing effects of small doses for a prolonged period (Coons, Leduc and Kaplan, 1951). These workers also used the fluorescent antibody technique to investigate the fate of oval-bumin, bovine serum albumin, and human γ-globulin following intravenous injection in mice. High concentrations of the antigens were detected within macrophages lining the venous sinuses of the spleen, and in 'reticulum cells' of the Billroth cords. Much less antigen was present in the cells of Malpighian bodies, and appeared to be present in lymphocytes and 'reticulum cells'. They also searched for antigen in regional lymph nodes after the injection of diphtheria toxoid into the foot-pad of rabbits, and located antigen in scattered cells in the medulla and in the cortex between lymphoid follicles (Leduc, Coons and Connolly, 1955). These studies demonstrated that the cells which took up the antigen did not themselves produce antibody. White (1963) also used the fluorescent antibody technique to study the fate of human serum albumin (HSA) in the spleen of chickens after intravenous injection, and found that, after 36 hours, most of the antigen was localized to certain cells, designated 'dendritic cells, probably a type of macrophage', in the germinal centres of the Malpighian bodies. The antigen seemed to be situated on the surface of the body and dendritic processes of the cells, and persisted at the site for at least two weeks, although it had meanwhile disappeared from macrophages in other areas of the organ. However, no antibody was demonstrable in the germinal centres during this period.

In an important series of experiments on antigen localization, Nossal and his colleagues initially used a different technique to trace the fate of injected antigens. They made extensive use of the flagellar antigen from *Salmonella adelaide*, which has the capacity of being highly antigenic, since as little as 0.0001 μg can induce antibody formation, and it can thus be used in very small doses. The advantage of using small doses in investigating the localization of injected antigen is that it avoids the possibility that, if large doses need to be given to ensure antibody formation, some of the antigen detected may represent the disposal of excess material rather than the localization of antigen involved in the immune response. The flagellar antigen was labelled with ^{131}I or ^{125}I, and the subsequent localization of the antigen was detected at the cellular level using autoradiographic techniques. When the labelled antigen was injected into the foot-pads of rats in doses not exceeding 10 μg, autoradiographs of sections and smears of the draining popliteal lymph nodes showed that heavily labelled macrophages were present in the medullary sinuses within 5 minutes of injection, and the cells of the subcapsular sinuses were similarly labelled after 30 minutes. Twenty-four hours after injection, the radioactive label was detectable in primary lymphoid follicles, where its pattern of distribution indicated its location in, and the presence of, cells having long dendritic processes extending between the lymphoid cells. These studies were extended to investigate the localization in popliteal lymph nodes of a wide variety of radio-iodine-labelled soluble and particulate antigens and 'non-antigenic' substances after injection into the rat foot-pad. It was found that all the antigenic substances localized to both medullary macrophages and lymphoid follicles as had been observed with flagellar antigens, whereas non-antigenic substances were taken up by medullary macrophages, but, with the exception of rat γ-globulin, did not localize to the lymphoid follicles. It was concluded that the differences in lymph node distribution of antigens and non-antigens were related to the ability of the dendritic macrophages to recognize 'foreignness' due to the association of the antigen with opsonizing factors (Nossal, Ada and Austin, 1964 *a, b*; Ada, Nossal and Pye, 1964).

Since the localization of ^{125}I by autoradiography in the above studies was not precise enough to show whether the label was in dendritic processes of phagocytes between the lymphoid cells, or actually within the lymphoid cells of the follicular region, Miller and Nossal (1964) combined the fluorescent antibody technique with studies of carbon uptake and silver impregnation to find the exact cellular localization in lymph nodes of the *Salmonella* flagellar antigen after injection into the foot-pad of the rat. In the primary response, antigen given in doses of 100 or 200 μg was followed by its localization in phagocytes of the medullary sinuses and in scattered dendritic cells in the diffuse cortex and medullary cords, but not in primary follicles. When the dose was raised to 2 mg, in addition to the features observed with the lower doses, a fine 'web' of stained material was seen irregularly distributed throughout the primary follicles. The pattern of distribution was similar to that which had been observed using radio-iodine-labelled antigen. At the end of the first day each web occupied most of the primary follicle, but, associated with the development of a germinal centre, by the fifth day eventually came to occupy a semilunar area capping the secondary follicle. During secondary immunization, follicular web staining patterns were seen irrespective of the dose of antigen. By the end of the first day and thereafter, caps composed of the characteristic webs of staining straddled the superficial margins of the germinal centres, lying both partially within, and partially above the germinal centres. Initially, deep in the germinal centres there were variable numbers of specifically stained cells, either round or dendritic, and by five days distinct webs appeared within the germinal centres themselves. Simultaneously, the caps became located wholly within the margins of the germinal centres. When antigen localization was studied following a previous non-specific stimulus, it was found that germinal centres *per se* were not necessarily more efficient in retaining antigen than primary follicles, but that increased retention in follicles at the time of a secondary stimulus was probably due to the opsonizing effect of previously induced specific antibody. Following the injection of colloidal carbon into the foot-pad, it was noted that there was a fine web of retained carbon in primary follicles and in germinal centres which closely resembled the pattern of antigen retention seen by the fluorescent staining, but the stain for metalophil cells or reticulin fibres did not reveal the same pattern. These authors concluded that the phagocytic cells responsible for the characteristic web pattern of antigen and carbon retention must be either functionally or morphologically distinct from the dendritic macrophages of the rest of the node. They also noted that the great surface area of the web would enable a very large number of lymphocytes to come into contact with antigen-retaining phagocytic cytoplasm, and observed that, once inside the web, a lymphocyte could not avoid contact with antigen on the surface of the phagocytes, or with phagocytes engaged in actively processing antigens. After intravenous injection of antigen, the antigen was initially located within macrophages in the red pulp, and was later located to dendritic cells within lymphoid follicles in a similar manner to that seen in lymph nodes after foot-pad injection (Nossal, Austin, Pye and Mitchell, 1966).

A similar distribution of antigen was found by McDevitt *et al.* (1966) who studied the localization of a synthetic polypeptide antigen labelled with ^{125}I in draining lymph nodes after its injection into the foot-pad in the mouse. In this study, they also identified antibody-producing cells by combining methyl-green–pyronin staining and immunofluorescent techniques. The lymph nodes were examined at intervals from 12 hours to 21 days. In unprimed mice, the label became predominantly located within macrophages of the medullary and cortical sinuses, but small amounts were present in germinal centres. In primed mice, label was rapidly and predominantly concentrated in germinal centres, where it persisted throughout the period of observation, while its intensity diminished in other areas. The pattern of distribution of label in germinal centres was web-like as previously described by Nossal and his co-workers. Specific antibody-producing cells did not at any time contain demonstrable antigen. The latter observations were in agreement with those of Nossal, Ada and Austin (1965) who teased lymph nodes into single cell suspensions following the injection of small amounts of ^{125}I-labelled flagellar antigen into the foot-pad of the rat. Single antibody-forming cells were identified and submitted individually to autoradiography. The findings indicated that the antibody-producing cells contained little or no macromolecular antigen, and it was considered unlikely that there were sufficient macro-molecules of antigen in a plasma cell to act as a direct template on polysomes for the formation of antibody.

Continuing their studies on antigen localization, Nossal and his colleagues proceeded to examine the subcellular localization of ^{125}I-labelled flagellar antigens using the technique of electron microscopic auto-radiography (Nossal, Abbot and Mitchell, 1968; Nossal, Abbot, Mitchell and Lummus, 1968). They examined the draining nodes after the injection of 20 μg of labelled antigen into the rat foot-pad as a primary or secondary stimulus. It was observed that antigen entered medullary macrophages by two ways: by apparent direct penetration of the plasma membrane, and by pinocytosis. In either case, the antigen became surrounded by tiny vesicles which may have represented Golgi-derived primary lysosomes. Vacuolar

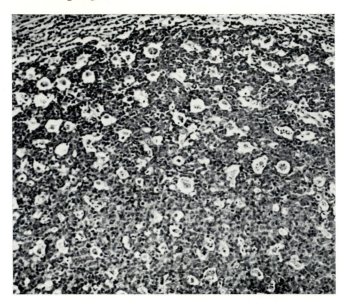

Fig. 45. 'Tingible body' macrophages scattered throughout germinal centre of lymph node follicle during secondary immune response. Haematoxylin–eosin. × 100.

fusion ensued and a series of progressively larger and more complex antigen-containing phagolysosomes were formed. The label persisted in such organelles for at least three weeks, and there was no evidence of the exit of labelled antigen fragments from the lysosomes during this period. No label was seen in plasma cells. The flagellar antigen was taken up vigorously and equally by both primary and secondary follicles, but the rate of uptake was faster in pre-immunized than in previously un-immunized animals. The bulk of antigen in the follicles persistently remained extracellular for at least three weeks. Label was found most frequently at or near the surface of fine cell processes, many of which were cytoplasmic processes of 'dendritic follicular reticular cells' (Fig. 47). Such processes interdigitated with equally fine but shorter processes of lympho-cytes, creating an elaborate meshwork. In some cases antigen was found between lymphocytes in close apposition, and there was some evidence suggesting entry of small amounts of antigen into lymphocytes. The characteristic 'tingible body' macrophages (TBM) of germinal centres (Figs. 45, 46) appeared to play only a secondary role in antigen retention, and they exhibited variable degrees of labelling over phagocytic inclusions. Follicles lacking in TBM retained antigen just as effectively as those with numerous TBM. However, larger numbers of TBM were consistently

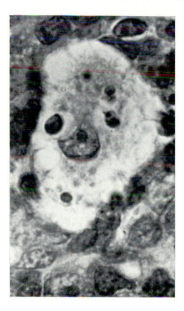

Fig. 46. 'Tingible body' macrophage showing abundant cytoplasm containing several dense bodies, and closely-packed surrounding lymphoid cells. Haematoxylin–eosin. × 1,200.

present in secondary follicles. It was concluded that the characteristic TBM of germinal centres appear to function primarily to phagocytose lymphocytes which have died locally, since they contain a large variety of inclusions and ingested nuclear material in various stages of breakdown. Why so many lymphocytes, most of which have undergone recent division, should die in germinal centres remains a mystery.

These studies indicate that, regardless of the mechanisms involved, follicular localization presents an extraordinary opportunity for contact of lymphocyte surfaces with antigen. The total surface area of the antigen-retaining mesh of reticular cell processes must be relatively enormous, and it is known that there is a significant traffic of recirculating lymphocytes through primary follicles. Antigen deposited in the follicles could cause some lymphocytes to undergo transformation and division, resulting in germinal centre formation.

From their studies on antigen localization, Nossal and his colleagues (Nossal, Abbot, Mitchell and Lummus, 1968) speculated that the following sequence of events may take place after antigen enters the tissues. Antigen enters the tissues where it may encounter natural or immune antibody; a soluble antigen–antibody complex is formed which is carried in lymph to the subcapsular sinus of a lymph node; some of the antigen–

antibody complex passes by the macrophages lining the sinuses and passes to the follicle where it becomes trapped on the antigen-retaining reticular cell. The trapped antigen then causes blast-cell transformation in antigen-reactive cells in constant motion through the follicular web of dendritic cell processes. Some of the transformed cells leave the follicle and move to medullary cords where they differentiate further to antibody-producing cells. Others remain in the follicle where they continue to divide and form the germinal centre. Antibody to the antigen is synthesized by cells in the medullary cords and red pulp of spleen, and is released into lymph and blood. The antibody circulates, and on reaching any unsaturated antigenic determinants retained in the follicular web, unites with them. The cycle is repeated and results in the accumulation of considerable quantities of immunoglobulins. This latter point is borne out by the demonstrable accumulation of antibody in germinal centres in repeatedly stimulated animals. As antigen entry into the node diminishes or ceases, antibody deposition on antigen, and antigen catabolism, terminate the inductive stimulus. Thus the rapid lymphocyte division ceases and the germinal centre regresses. Renewed antigen supplies may subsequently initiate another similar cycle.

Somewhat similar hypotheses were put forward by White, French and Stark (1970) on the basis of their findings in the spleen of the chicken. Chickens which had been given an intravenous injection of human serum albumin (HSA) produced a rapid rise of antibody to a peak at 8–9 days. Plasma cells containing anti-HSA appeared in the spleen at 24 hours after injection. They produced evidence that early synthesized antibody causes the localization of antigen (HSA), presumably as complexes, to cells in the white pulp of the spleen at the periphery of ellipsoids (Schweigger–Seidel sheaths) by 24–32 hours. Inspection of a series of tissue sections showed that antigen-bearing cells appear to migrate through the white pulp and subsequently to appear (88 hours–6 days) as dendritic cells within germinal centres, which originated in the angle between the diverging penicillary arterioles at their point of origin from the central arteriole of the white pulp. It was postulated that germinal centre formation ensues by a process of progressive capture and aggregation of lymphocytes at the surface of antigen-bearing dendritic cells, being an agglutinative reaction between the antigen-bearing cells and small lymphocytes with antibody at their surface. The reaction between antigen and antibody on the surface of the lymphocytes would be the stimulus for the pyroninophilic transformation and subsequent mitotic division which is seen in early germinal centres. The prolonged persistence of antigen (about 42 days) in or on the dendritic cells in close proximity to the dividing lymphoid cells in the germinal centres, indicates a significant role for this small proportion of antigen in directing lymphocyte proliferation. The lymphocytes initially retained within the germinal centre are demonstrably immunologically competent, and therefore their specificity must, to explain an agglutinative reaction, correspond to that of the injected antigen. The proliferation of these retained lymphocytes would probably result in offspring of similar specificity, and would provide the means for an increase in avidity of antibody synthesized later in the immune response or following a secondary stimulus.

In summary, the findings quoted above show that subcutaneously or intravenously injected antigens are primarily taken up by macrophages in lymph nodes or spleen respectively. At a later stage, a proportion of the antigen becomes localized to the plasma membrane of the dendritic processes of antigen-retaining reticular cells in lymphoid follicles in lymph nodes or spleen. There is no clear evidence that antigen is taken up by antibody-producing cells.

It is apparent that the 'antigen-retaining reticular cell' seems to play a key role in the immunological response by forming a meshwork in lymphoid follicles in which recirculating lymphocytes may be selectively 'trapped' by an agglutinative reaction between the antigen on the surface of the reticular cells and specific antibody present on the surfaces of the lymphocytes. The subsequent division and proliferation of trapped lymphocytes could give rise to the appearance of germinal centres. The obvious question arises as to whether the antigen-retaining reticular cell is a unique type of cell, or a specialized type of macrophage. Opinions differ upon this point. The reticular cell is characterized by the presence of numerous long processes which are narrow even at their point of origin from the cell, measuring from $100 \, \text{m}\mu$ to 2μ in width. There is a relative paucity of dense granules and phagocytic vacuoles, and fewer mitochondria than in normal macrophages. The cytoplasm contains numerous Golgi-type vesicles and occasional strands of RER. Microtubules and microfilaments are present. The processes of adjacent reticular cells may be bound by desmosomes, particularly as germinal centres reach their maximum size. They differ markedly from the other macrophages of lymphoid follicles and germinal centres, which, although often having some tapering dendritic cell processes, usually contain many assorted dense bodies which have resulted from endocytic activity. Thus it is evident that the antigen-retaining reticular cell differs morphologically from classical macrophages; and, since the former cell also appears to behave in a unique way during immunological responses, it would seem to be sensible to distinguish it, by name, from the latter until the relationship between the two types of cell is clarified.

Recently, Fischer *et al.* (1970) have described the localization of carbon and antigens to amoeboid macrophages and dendritic phagocytes in the omentum; the lack of demonstrable metalophilia in the dendritic processes of the latter cells was in agreement with the findings of Miller and Nossal (1964) in antigen-retaining reticular cells of lymph nodes. However, Fischer and his associates were of the opinion that the dendritic cells could migrate from the omentum after transforming into amoeboid macrophages. The dendritic phagocytes were also found to proliferate during antigen stimulation, but irradiation reduced their capacity to localize antigen and to interact with lymphoid cells. The authors speculated that the dendritic cells may arise from primitive non-phagocytic reticulum cells, and that the dendritic cells themselves represent an intermediate stage of differentiation tending towards the wandering peritoneal macrophage.

THE SUBCELLULAR DISTRIBUTION OF ANTIGENS

The subcellular distribution of injected antigen has also been studied using a number of techniques. Using the technique of electron microscopic autoradiography, Nossal, Abbot and Mitchell (1968) showed that ^{125}I-labelled flagellar antigen was rapidly endocytosed and segregated into phagolysosomes within medullary macrophages in lymph nodes, and there was no evidence of the release of labelled material during the three weeks of the experiment. A proportion of the antigen later became associated with the surfaces of the processes of the antigen-retaining reticular cells of germinal centres, where it apparently remained for a prolonged period. Using a similar technique, Szakel and Hanna (1968) investigated the localization of ^{125}I-labelled human gamma globulin (HGG) in the spleen during the primary immune response following the intravenous injection of the antigen in mice. They found that the label became localized to areas of numerous infoldings of the plasma membranes of the antigen-retaining reticular cells: the infoldings were such that the cell membrane constituting opposing walls of channels formed by infolding were parallel and narrowly separated by extensions of the extracellular space. The development of the infoldings appeared to be antigen-related and increased with time, and after day 4 the plasma membranes of the antigen-retaining reticular cells were closely associated with cytoplasmic processes of immunoblasts in some places. During the secondary response to HGG (Hanna and Szakel, 1968) the label became located over the infoldings of antigen-retaining reticular cells within 2 hours of injection (Fig. 47). Small villous extensions of the cytoplasm of immunoblasts were closely associated with the infoldings in some areas, and numerous morphologically similar vesicles were observed within the cytoplasm of both types of cell. From day 1 onwards, there was progressive loss of regularity and organization of the infoldings of the reticular cells, so that the walls were no longer parallel, and electron-dense material formed pools between the

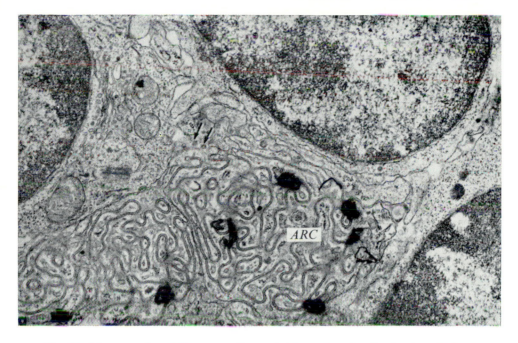

Fig. 47. Electron microscope autoradiograph illustrating the distribution of silver grains over the labyrinthine plasma membrane infoldings of antigen-retaining reticular cell processes (*ARC*) 2 hours after the initiation of the secondary immune response in mouse spleen germinal centres with ^{125}I-labelled human γ-globulin. The membrane infoldings are in close apposition to surrounding immunoblasts and small vesicles (arrows) are in close proximity to the labelled cell process. × 26,300. Reproduced with permission from Hanna and Szakel (1968).

walls of infoldings. The disorganization of the infoldings progressed, with the formation of larger pools of electron-dense material, and disruption of the plasma membrane could be observed from the third day onwards. By day 7, disorganized remnants of the reticular cells were present, and there was a marked increase in the number of phagocytic inclusions within the tingible body macrophages. During these studies, some of the label became localized in the region of lysosomes within macrophages in the germinal centres.

Other investigators have studied the distribution of antigen in various fractions of cell and tissue homogenates. Thus it has been shown that after the injection of various radio-iodine-labelled antigens into the rat foot-pad, in homogenates of draining lymph nodes the label was associated with the 'large granule' subcellular fraction which was membrane bound, and therefore probably within lysosomes or phagolysosomes, and was still associated with the antigen (Ada and Lang, 1966; Ada and Williams, 1966). Similar results were obtained by Uhr and Weissmann (1965) who injected bacteriophage T_2 or $\Phi X174$ intravenously in the rat and guinea-pig and subsequently found the phage associated with the large granule fraction of liver homogenates from which it could be released by lyso-lecithin, which disrupts membranes, and was therefore presumably located within the lysosomes of Kupffer cells. Kölsch and Mitchison (1968) studied the fate of phagocytosed antigens in cells from peritoneal exudates in mice using ^{125}I- and ^{131}I-labelled antigens. After uptake of antigen, the cells were homogenized by isopycnic centrifugation in a sucrose gradient. Ninety per cent of heat-denatured BSA, BSA or BGG was found in the cellular compartment showing the characteristics of lysosomes, and was rapidly degraded in this compartment, having a half-life of less than one hour. Storage of antigen occurred in a separate com-partment, the nature of which was not clear but possibly associated with the nucleus in some way. Most of the antigen in the latter compartment was irreversibly attached and protected from degradation.

Taken as a whole, the above studies indicate that a proportion of injected antigen is retained in unchanged form on the surface of antigen-retaining reticular cells and plays a part in immune induction, whereas the antigen taken up by macrophages is probably sequestered within phago-lysosomes where it may or may not undergo some form of degradation; and on the basis of the evidence hitherto presented, would not, on morphological grounds alone, appear to play a part in immune induction.

A variety of experimental systems have shown that live macrophages previously exposed to antigen can interact with syngeneic lymphoid cells and elicit an antibody response *in vivo* or *in vitro*. Thus when rat peritoneal macrophages, 20 minutes after the uptake of lysed sheep red blood cells (SRBC) are allowed to interact with thoracic duct lymphocytes *in vitro* for several hours, and the lymphocytes are subsequently separated from the macrophages and transferred to irradiated hosts, the recipients show an antibody response within a few days. Thoracic duct lymphocytes incu-bated *in vitro* in the absence of macrophages do not elicit antibody forma-tion when transferred to irradiated recipients (Ford, Gowans and McCullagh, 1966). Moreover, peritoneal macrophages exposed to antigen and transferred to syngeneic recipients are far more effective in priming the animals for a secondary response than equivalent amounts of soluble antigen given as a single injection. Heterologous macrophages are far less effective than isologous ones since live macrophages result in a better response than dead ones, and presumably heterologous cells are rapidly eliminated from the recipient (Askonas, Auzins and Unanue, 1968). These findings, which are in general agreement with those of other investigators, indicate that the role of macrophages following antigen injection may not be entirely restricted to acting as scavengers in taking up and degrading the foreign material. The fact that there is hitherto no evidence that antibody-producing cells take up significant amounts of antigen, whereas antigen-retaining reticular cells and macrophages undoubtedly take up and retain antigen, indicates that antibody-producing cells may receive some form of immunogenic 'message' from cells which take up antigen.

The evidence which supports the concept of antigen retained on the surfaces of the plasma membranes of the antigen-retaining reticular cells (which may be specialized macrophages) is largely morphological in that the cells have very close contact with immunoblasts in germinal centres. The fact that macrophages previously exposed to antigen are able to initiate the production of specific antibody by unprimed lymphoid cells, whereas antigen exposed to unprimed lymphoid cells in the absence of macrophages cannot elicit antibody production, indicates that morphologically unspecialized macrophages are capable of 'handling' the antigen in such a way that it is effectively immunogenic. Elucidation of the role of macrophages in immune responses is largely dependent upon the experimental elicitation of antibody formation *in vitro*, and some caution must be used in directly relating the results to the situation which obtains during immune responses *in vivo*. However, the findings warrant a closer examination of the fate and immunogenicity of antigen following its uptake by macrophages.

THE IMMUNOGENICITY OF ANTIGEN FOLLOWING UPTAKE BY MACROPHAGES

In essence, the previously cited evidence indicates that, since macrophages take up antigen and antibody-producing cells do not, it seems logical to postulate that the latter must receive some form of immunogenic stimulus or 'message' from the former. The extensive studies of Fishman and his colleagues are highly relevant to this theory, and the results of their early experiments have been summarized by Fishman and Adler (1963). They found that cultures of immunologically-competent cells alone did not produce antibody when exposed for the first time to antigen; but that such cultures did produce specific antibody against T_2 bacteriophage when a cell-free filtrate, prepared by exposing the T_2 bacteriophage to peritoneal macrophages *in vitro*, was added to lymph node cells. Specific antibody was not produced by similar cultures exposed directly to T_2 bacteriophage, or to macrophages which had not been in contact with the antigen. The maximum antigenicity of the macrophage extracts was achieved after 30 minutes of incubation of the macrophages with the antigen, and 10 minutes of exposure yielded an inactive extract. Finally, it was demonstrated that the formation of antibody against T_2 bacteriophage was not induced in this system when the macrophage extracts were heated to 80–100 °C for 15 minutes or when ribonuclease (RNAse) was added to the lymph node culture medium, whereas deoxyribonuclease (DNAse) or proteolytic enzymes had no effect. These observations suggested that the cell-free extract of macrophages which stimulated antibody synthesis by the lymphoid cells was RNA, and led Fishman and Adler to investigate the immunogenicity of RNA extracted from macrophages after exposure to antigen. In these experiments they used X-irradiated rats in which were implanted diffusion chambers containing normal rat lymph node cells: the rats were themselves unable to produce antibody following the irradiation, and had low levels of natural phage neutralizing antibody. Significant antibody production occurred in the majority of the recipients bearing chambers containing lymph node cells and cell-free extracts of macrophages incubated with T_2 bacteriophage, or containing lymph node cells and RNA extracted from macrophages incubated with T_2 bacteriophage; but not in the recipients with chambers containing lymph node cells plus RNA from T_2-stimulated macrophages plus RNAse, or containing lymph node cells plus RNA from unstimulated macrophages plus T_2 bacteriophage, or RNA from T_2-stimulated macrophages without the presence of lymph node cells. In these studies, the antibody produced was of the 19S (IgM) type, and the RNA extracted by the phenol method from heated macrophages was a crude preparation obtained from macrophages incubated with very small amounts of T_2 bacteriophage (1 plaque-forming unit per 1,000 macrophages), since larger doses of antigen were less effective. In a later study, Fishman, van Rood and Adler (1965) found that higher doses of T_2 bacteriophage (up to 100 plaque-forming units per macrophage) were effective when the extraction procedure yielded a more pure RNA preparation. Moreover, the RNA prepared in this way induced two waves of antibody formation when added to fragments of lymph nodes from normal rats; the first wave comprised 19S (IgM) antibodies, and the second wave 7S (IgG) antibodies. However, the earlier IgM antibodies were later found to possess allotypic markers characteristic of the rabbits serving as donors for peritoneal cells and RNA, whereas the later IgG antibody had the allotypic specificity of the donor of the lymphoid cells (Adler, Fishman and Dray, 1966). The findings of Fishman and his colleagues led them to postulate that the uptake of antigen by macrophages leads to the formation of a low molecular weight RNA, and that this RNA is taken up by lymphoid cells which are then stimulated to produce specific antibody.

The findings of Fishman and his group have received wide support from the observations of many other workers. Attempts have also been made to see whether the stimulus to antibody formation is constituted by RNA alone or RNA combined with fragments of antigen, since the majority of evidence excludes intact antigen from forming part of the immunogenic message. In this connection, Askonas and Rhodes (1965) exposed ^{131}I-labelled haemocyanin (HCY) to mouse peritoneal macrophages and subsequently extracted their RNA; the extract was then incubated with mouse spleen cells, which were then injected intraperitoneally into mice previously primed with the same antigen. RNA extracts made immediately following exposure to antigen were immunogenic, indicating that immunogenicity was not due to the synthesis of new RNA by macrophages following contact with antigen; moreover, RNA extracted after $2\frac{1}{2}$ hours of exposure to antigen was not only antigenic but also contained ^{131}I-labelled material. The antibody response was inhibited by treatment of the extracts with RNAse, and it was postulated that macrophage RNA may either facilitate antigen uptake by macrophages, or protect antigenic determinants from degradation following the ingestion of antigen. Similarly, Friedman, Stavitsky and Solomon (1965) showed that extracts of RNA from macrophages incubated with T_2 bacteriophage also contained phage antigens detectable by complement fixation.

Others have also shown that RNA–antigen complexes can be extracted from macrophages after exposure to antigen, and the question arises as to whether (1) these RNA–antigen complexes are a product of the extraction procedures used; (2) whether it is essential for antigen to be complexed to RNA for it to be an effective stimulus to antibody production; (3) whether the complexing of antigen to RNA protects the antigen from degradation in the phagosome–lysosome system of the macrophage; or (4) whether the RNA purely functions in an adjuvant-type capacity. Highly relevant to these possibilities are the recent findings of Roelants and Goodman (1969), who investigated various antigens, the ribonucleic acids, the involvement of specific enzymes, and the requirement of a specific type of cell in relation to the possible role of macrophage RNA–antigen complexes in immune induction. They found that the formation of RNA–antigen complexes *in vitro* was unrelated to the immunopotency of the antigen, was not followed by an enzyme-dependent reaction, did not require the synthesis of RNA following introduction of the antigen, did not seem to involve antigen-specific ribonucleic acids, was not specific for macrophages since HeLa cells could be used as effectively, and occurred when purified RNA was mixed with antigen only in the presence of divalent cations. The complexes were very stable once formed, but could be dissociated by exhaustive dialysis against buffers containing a chelating agent. It was concluded that the macrophage–RNA complex appeared to be a chelate between anionic groups on the two components. Based on the total absence of a relationship between immunogenicity and the capacity to form such complexes at every level examined, it was concluded that it is unlikely that RNA–antigen complexes play a physiologically significant part in immune induction. Moreover, it has been shown by others (see Askonas *et al.* 1968) that RNA does not have to be newly formed within macrophages on contact with antigen for the stimulation of antibody synthesis to be effective, and that in the presence of excess antigen, RNA–antigen complexes can be extracted from suspensions of macrophages in the absence of antigen uptake; also, RNA extracts of macrophages not having had contact with antigen contain a similar RNA complexed with an unknown protein.

These findings would appear to exclude macrophage RNA from having an essential role in informative transfer. The possibility still remains that, if new antibody is produced by a clonal selection mechanism, then the responsive cells could conceivably be 'triggered' by a messenger RNA and need not receive a completely new coding message, although there are numerous convincing arguments against such a role of a messenger RNA which codes for specificity and allotypic sequences of immunoglobulins (Askonas *et al.* 1968). Recently, Fishman and Adler (1970) have reviewed the evidence which they interpret as indicating that macrophages are capable of producing highly immunogenic complexes of antigen and RNA, although they recognize that not all RNA–antigen complexes are necessarily immunogenic. They suggest that the complexes are products of a distinct sub-class of macrophages since peritoneal macrophages synthesized 'immunogenic' RNA after exposure to antigen, whereas alveolar macrophages were inactive in this respect.

It has recently been demonstrated in several laboratories that a proportion of antigen becomes located in the region of the plasma membrane of the macrophage and is retained there for a period of time without undergoing digestion in the phagosome–lysosome system of the cell. Thus Unanue and Askonas (1968) attempted to correlate the uptake, catabolism, persistence and immunogenicity of antigen in macrophages. Peritoneal macrophages were cultured for several hours after uptake of ^{131}I-labelled haemocyanin (HCY), and it was found that the cells degraded most of the protein within 2–5 hours. However, the ability of the cells to prime lymphocytes of syngeneic mice for a secondary response remained unchanged for at least two weeks despite the loss of more than 90 per cent of their original content of antigen. The persistence of immunogenicity was associated with a small percentage (about 10 per cent) of antigen retained by the cell in a form which was protected from rapid breakdown and elimination. The stable antigen appeared to be localized in an area or compartment of the cell not subjected to normal digestive processes, and also accessible to contact with the immunologically competent cell. Whether the antigen passed through a phagolysosomal stage before being sequestered in a stable form was not determined. Radioautography at the ultrastructural level revealed that a large part of the retained antigen was in close association with the plasma membrane of the cell. Further studies by Unanue and Cerottini (1970) confirmed that most of the ^{131}I-labelled HCY was rapidly catabolized and eliminated by cultured macrophages, and that a few molecules of HCY were retained on the plasma membrane of the cells for prolonged periods and were not subjected to endocytosis or catabolism. The membrane-bound antigen, which could be removed by trypsin or EDTA, was of large molecular size but heterogeneous. A large part of the immune response of mice to HCY bound to live macrophages could be abrogated by prior treatment of the macrophages *in vitro* with antibody or trypsin. Hence, most of the immunogenicity of HCY bound to macrophages was attributed to the few molecules of antigen bound to the plasma membrane.

Broadly similar results were obtained by Mitchison (1969) who compared the immunogenic capacity of various protein antigens (lysozyme, ovalbumin, HSA and BSA) in the free and peritoneal macrophage-bound forms. The cell-bound antigen was far more potent than the free form in inducing primary immunization. It was also found that (1) viable macrophages were required; (2) irradiation of the cell donors 2–7 days before giving antigen inhibited immunization, but irradiation after uptake did not do so; (3) macrophages do not retain large amounts of antigen for long; (4) the immunogenic activity did not depend solely on a minor phagocytosis-prone fraction of the antigen; (5) immunologically paralysed hosts are not susceptible to immunization, but macrophages from paralysed donors are effective; and (6) the enhancement of immunogenic capacity does not apply to the secondary response. It was concluded that the findings did not support the concept of an RNA message nor of a 'superantigen' produced by macrophages linking antigen to RNA. The ability of macrophage-bound BSA to immunize mice during the phase of recovery from paralysis, while no concentration of free BSA would immunize under the same circumstances, was taken to indicate that macrophages do not merely serve as an antibody-concentrating mechanism.

The above findings suggest various possibilities for the immunogenic functions of macrophages. Thus, macrophages could carry a small percentage of antigen, in relatively stable form and isolated from catabolic processes, to foci of immunologically competent lymphoid cells, and the superior immunogenicity of antigen bound to macrophages relative to free soluble antigen may relate to the greater opportunity for the cell-membrane-bound antigen to interact with lymphoid cells. There is obviously a striking resemblance here to the role envisaged for the antigen-retaining reticular cells in lymphoid follicles.

Recent evidence that the precursor of the antibody-synthesizing lymphoid cell, derived from the bone marrow, is induced to differentiate into an antibody-forming cell by an antigen-reactive cell, also of lymphoid type but derived from the thymus (Miller and Mitchell, 1968; Mitchell and Miller, 1968; Nossal, Cunningham, Mitchell and Miller, 1968), is in keeping with the concept of macrophages exposing unprocessed membrane-located antigen to lymphoid cells in a fashion which is more effective than in the case of free antigen, and tells against the hypothesis that there is transfer of processed antigen, RNA–antigen complex or a messenger RNA from macrophages directly to immunologically competent cells. Although lymphocytes appear to be genetically endowed with the capability of forming certain immunoglobulin molecules, they seem to require aid in reacting with and responding directly to some antigens. Perhaps the macrophage-bound antigen stimulates the proliferation and differentiation of lymphoid cells, and immunological memory could be accounted for by the membrane-bound antigen being slowly released from the macrophages to stimulate the persistence of primed lymphoid 'memory' cells (Unanue and Askonas, 1968).

It would thus appear that, under experimental conditions, less than 10 per cent of antigen is retained on the plasma membrane of macrophages, and the remainder is endocytosed and degraded in phagolysosomes. The degradation of most of the antigen is consistent with the role of macrophages in endocytosing and digesting foreign material. It is not known whether the antigen which is retained on the plasma membrane is located in a relatively small number of special receptor sites, or why its fate should differ from that of the remainder of the foreign material. It is not known whether antigen binding to cell membranes is an active process or a haphazard event, but the possibility that it may involve membrane RNA could account for the antigen–RNA complexes isolated from macrophages by Fishman and Adler (1970).

Since endocytosis of antigen would not now appear to play a part in the immunogenic activity of macrophages, then factors which stimulate or depress the phagocytic activity of macrophages would not appear to be operable in directly stimulating or depressing antibody production. The fact that stimulation or depression of the phagocytic activity of macrophages *in vivo* is in many cases accompanied by stimulation or depression of antibody production (see Nelson, 1969), lends support to the concept of macrophages normally being involved in the immune response; but it also suggests that increased or reduced endocytosis may be paralleled by changes in the amount of antigen bound to the cell membrane, rather than alterations in the processing of antigen as has been suggested previously. Moreover, changes in antibody production which may accompany experimentally induced changes in the phagocytic activity of macrophages cannot necessarily be interpreted as being directly related phenomena. For example, corticosteroids depress phagocytic activity and the immune response, but also cause depletion of lymphoid tissues.

During the secondary response, antibodies are produced more rapidly and to higher titres than in the primary response. This may be attributable to a direct or indirect effect of antigen on cells already committed to, but not actually engaged in, the production of specific antibody. The responding cells, referred to as 'memory cells', are generally considered to have the morphology of small lymphocytes, and the ability to mount a secondary response is referred to as 'immunological memory'. The obvious question arises as to whether antigen is directly effective in stimulating the memory cells, or whether it must first be processed or membrane-bound in macrophages. Of course, as in the case of the primary response, it is not possible to exclude completely the possible localization of undetected antigen fragments within antibody-producing cells, or the possible stimulation of such cells by transient contact with intact antigen. Although there is not uniform agreement on this matter, the majority of experimental findings support the view that secondarily injected antigen does not localize within antibody-producing cells; but there is universal agreement that it becomes localized to macrophages and antigen-retaining reticular cells. It has been reported that antigen retained on the cell membrane of peritoneal macrophages is not more potent than free antigen during secondary responses, whereas the cell-bound antigen is much more potent during primary responses (Mitchison, 1969). This suggests that free antigen may be effective in directly stimulating the division of memory cells which have been set aside as a result of a select population of lymphoid cells receiving an initial stimulus from cell-bound antigen. Alternatively, antibody production during secondary responses could be accelerated by the rapid release of stored cell-bound antigen (or a messenger RNA) after secondary contact with antigen.

INTERACTION BETWEEN MACROPHAGES AND LYMPHOCYTES

The overwhelming majority of the evidence cited above indicates that the localization of antigen to macrophages is an essential part of the primary immune response, and of necessity invokes the concept of the macrophages subsequently interacting with antigen-reactive lymphoid cells in such a way that exposure or transfer of antigen or other messenger material takes place. There is evidence that the micro-anatomical relationship between macrophages and lymphoid cells is such that there is ample opportunity for the transformation of immunogenic information to take place both *in vivo* and *in vitro*.

The occurrence of 'peripolesis' (the congregation of lymphocytes around macrophages) in cultures of lymph nodes regional to skin allografts and of spleen after intravenous antigen injection has been recorded by cinemicrography (Sharp and Burwell, 1960). In similar studies, Berman (1966) examined long-term cultures of lymphocytes and macrophages, and observed that the lymphocytes in peripolesis around macrophages appeared to project pseudopodia which indented or penetrated the cell membranes of the macrophages. McFarland and Heilman (1965) observed that in mixed leucocyte cultures from unrelated human donors, on the second day, lymphocytes became firmly attached to macrophages by means of a single cytoplasmic process or 'foot appendage', and as many as ten lymphocytes could be firmly attached in radial fashion around the macrophage in this way. By the third day, some of the lymphocytes began to undergo the transformation into larger blast-type cells which is normally observed in mixed leucocyte cultures of this sort. There are also many other reports of small lymphocyte-like cells having a close spatial relationship *in vitro* with the surfaces of other cells, many of which would appear to be macrophages (see Sharp, 1968). The existence of this phenomenon *in vitro* does not mean that this situation necessarily occurs as a dynamic situation *in vivo*; however, several investigators have described the close association of macrophages and lymphoid cells during immune responses *in vivo*.

In electron microscopic studies, Sorensen (1960) and Han (1961) described the presence of clusters of lymphocytes or plasma cells, or combinations of both cell types, around macrophages in the medullary cords of lymph nodes. These findings were supported by the observations of Schoenberg, Mumaw, Moore and Weisberger (1964) who also observed clusters of lymphoid cells surrounding macrophages in the medullary cords of lymph nodes and in the red pulp of the spleen in the rabbit. These clusters were more numerous in immunized animals. They also observed infrequent areas of distinct communication between the cytoplasm of the macrophages and some of the immediately adjacent lymphocytes and plasma cells, and within these areas of cytoplasmic fusion particles having the appearance of ribosomes were observed. The transfer of antigenic molecules between macrophages and lymphocytes was not seen when horse ferritin was used as an electron-dense antigen. They concluded that ribosomal particles may pass from one cell to another by means of cytoplasmic bridges. Others have similarly suggested that processed antigen, RNA, or antigen–RNA complexes may pass from macrophages to lymphoid cells by means of cytoplasmic bridges, but there is no satisfactory evidence that such interchange takes place.

Since present evidence indicates that macrophages are involved in the immunogenic 'exposing' of unprocessed cell-membrane-bound antigen to lymphoid cells, then it is reasonable to assume that this would certainly require intimate contact, and possibly cytoplasmic fusion, of the cell membranes of the two cell types. The possible importance of the close micro-anatomical relationship between antigen-retaining reticular cells and lymphoid cells in germinal centres of spleen and lymph nodes has been discussed previously.

THE DIVERSITY OF IMMUNOLOGICAL RESPONSES

This account of the role of the macrophages in immunological responses has hitherto been confined largely to those events which lead to antibody production in its widest sense, and has not taken into account the diversity of immunological responses normally seen in mammals. A distinction is usually drawn between *humoral* immunity which is due to the presence of circulating specific antibodies, and *cellular* immunity which is a state of altered reactivity not due to demonstrable circulating antibodies but to antibodies bound to living cells. Various aspects of cellular immunity are dealt with in the preceding chapter.

The serum globulins which exhibit antibody activity are now designated as immunoglobulins or gammaglobulins; five classes are distinguished at the present time on the basis of various characteristics, and are usually designated as IgG, IgA, IgM, IgD and IgE. In the newborn animal, the first antibodies which appear belong to the 19S (IgM) category. During primary responses, the initial humoral antibody synthesized is of the 19S type, but, after a period of time which may be some weeks in newborn mammals or a few days in mature mammals, there follows the production of 7S antibody which is mainly of the IgG type; and the production of 7S antibody may subsequently proceed at a low level for a very long period of time. During the secondary response, predominantly 7S antibodies are produced, mainly of the IgG type. However, some antigens will continue to elicit the production of antibodies belonging to the 19S fraction even during secondary exposure. It is still unclear as to whether 19S and 7S antibodies are produced sequentially by the same individual cells, or by different cells.

Fishman *et al.* (1965) and Adler *et al.* (1966) found that RNA extracted from rabbit macrophages exposed to T_2 bacteriophage *in vitro* induced two waves of antibody formation when added to fragments of rat lymph nodes in culture, and that the earlier IgM antibodies had the allotypic markers characteristic of the rabbit donors whereas the later IgG antibodies had the allotypic specificity of the rat donors of the lymph node cells. This suggests that, in this system, the immunogenic information inducing antibody synthesis was different from that inducing the later IgG antibody synthesis. Moreover, the RNA extract inducing the earlier IgM antibody synthesis was more susceptible to treatment with ribonuclease than that which was presumed to stimulate the later IgG antibody synthesis. Incubation of the macrophages with actinomycin D before contact with bacteriophage did not affect the production of the later IgG antibody, although the earlier IgM antibody response appeared to be abolished. These findings support the concept that the formation of IgM antibodies does not govern the subsequent formation of IgG antibodies, and it is tempting to postulate that macrophages are more concerned with the sequence of events culminating in IgM antibody formation, rather than the subsequent formation of IgG antibodies.

It appears certain that not all antigens need to follow the same pathway in eliciting an immune response. Thus a recent study has indicated that, in the mouse, macrophages are not required in initiating an immune response against polymerized bacterial flagellin, but are necessary for mounting a response against sheep erythrocytes (Shortman, Diener, Russell and Armstrong, 1970). It remains to be seen whether the differences in cellular requirements reported are as absolute as the data suggested, but it would be of interest to know the relative proportions of IgM and IgG antibodies in each of the responses. Although present evidence suggests that lymphocytes are genetically capable of forming certain immunoglobulin molecules, they appear to require the intervention of macrophages in responding to weak or soluble antigens during primary immunization at least, but it seems quite possible that other antigens may stimulate antigen-reactive lymphocytes without the necessity for the intervention of macrophages. Conversely, the uptake of antigen by macrophages does not always ensure high immunogenicity (Askonas *et al.* 1968). There have also been reports that the uptake of sheep erythrocytes decreases their immunogenicity (Perkins and Makinodan, 1965), although recent studies suggest that this observation was related to the relative proportions of cells used (Shortman *et al.* 1970).

Studies on immunological tolerance (paralysis) have revealed some information on the involvement of macrophages in altered immune responses. Immunological tolerance is the state which exists when there is a reduction or complete loss of ability to mount an immune response against a particular antigen, without affecting the ability to react to other antigens. Martin (1966) demonstrated that the majority of neonatal rabbits injected with BSA responded poorly to subsequent challenge with the same antigen; whereas the majority of neonatal rabbits injected with BSA plus peritoneal macrophages from adult rabbits responded to subsequent challenge with BSA by producing demonstrable precipitating antibody. Mitchison (1969) found that macrophage-bound BSA can immunize mice during the phase of recovery from tolerance, whereas no amount of free BSA would immunize under the same circumstances. These findings suggest that macrophages do not merely act as an antibody-concentrating mechanism during immune responses.

Although immunologically tolerant hosts are not susceptible to immunization, macrophages from tolerant donors are just as effective as normal macrophages in priming recipients (Humphrey and Frank, 1967; Mitchison, 1969), and can stimulate DNA synthesis of primed spleen cells *in vitro* as efficiently as normal macrophages (Harris, 1966). Moreover, it has been shown that labelled HSA and HCY are taken up to the same extent in tolerant rabbits and in normal adult rabbits (Humphrey and Frank, 1967), whereas antigen injected into neonatal rats to produce tolerance exhibits little or no uptake by macrophages in the recipient animals (Mitchell and Nossal, 1966).

It is widely recognized that tolerance may be induced by injecting large amounts of antigen into newborn, irradiated adult, or normal adult animals, although the latter require much higher doses of the same antigen than the first two groups. In other cases, tolerance may be induced by the same antigen either administered in very low doses or in very high doses, whereas in the intermediate range of doses normal antibody formation occurs. In experiments on tolerance to bovine IgG (BGG) in adult mice, Dresser (1961, 1962) found that when aggregated BGG was removed by ultracentrifugation, the remaining soluble fraction of BGG was not immunogenic; however, the soluble fraction became immunogenic when emulsified with Freund's complete adjuvant, whereas the aggregated fraction of BGG was immunogenic without necessitating emulsification. In contrast, it was easy to induce tolerance with very small amounts of the soluble fraction of BGG, and this tolerance could not be overcome by mixing the antigen with adjuvant. Frei, Benacerraf and Thorbecke (1965) recovered a non-phagocytozed fraction of ^{131}I-labelled BSA from the serum of rabbits after the intravenous injection of ^{131}I-BSA. They injected theoretically immunogenic quantities of the recovered non-phagocytozed BSA into normal rabbits and found that it either produced tolerance to BSA, or a very poor antibody response when compared with the original ^{131}I-BSA.

A number of hypotheses exist with regard to the mechanisms underlying immunological tolerance. On the basis of the above experimental findings, it can be postulated that, under normal circumstances, antigen becomes localized to macrophages and subsequently gives rise to a normal immunological response; however, if antigen escapes uptake by macrophages, either because the macrophages may be deficient themselves as in neonatal animals, or because the antigen is in a form which is not taken up by the cells, or because there is too little or too much antigen, then the free antigen could act directly on immunologically-competent cells to produce tolerance. That recognition and uptake of antigen by macrophages are not impaired in tolerant adult animals is consistent with current theories that tolerance is due to the elimination of antigen-reactive lymphoid cells by excess antigen.

Little is known of the possible role of macrophages in the induction of 'cell-bound' antibody formation. Macrophages participate in many of the events associated with the state of delayed-type hypersensitivity, and delayed-type hypersensitivity reactions are expressed during cell-mediated antibacterial and anticellular immunity. The production of significant amounts of cytophilic antibodies in guinea-pigs requires the incorporation of the antigen into Freund's complete adjuvant; and this is also required for the induction of unequivocal and lasting delayed-type hypersensitivity (Nelson and Boyden, 1967). Delayed-type hypersensitivity and humoral antibody formation are two distinct immune reactions which can exhibit apparently identical specificity for antigen, which suggest that the specificity of delayed-type hypersensitivity could be due to an essential participating role of antibody. Delayed-type hypersensitivity can be demonstrated in an animal well before the formation of humoral antibodies, and may also be demonstrated in animals both agammaglobulinaemic and totally incapable of eliciting detectable antibody to any antigen. In contrast, procedures such as neonatal thymectomy, administration of anti-lymphocyte serum or 6-mercaptopurine, or pre-treatment with antigen, can depress or abolish delayed-type hypersensitivity without affecting humoral antibody formation (Szenberg and Warner, 1967). In many situations, delayed-type hypersensitivity and classical antibody production take place concurrently in response to antigen administration. However, sensitization involving the predominant or exclusive production of delayed-type hypersensitivity differs morphologically from that of humoral antibody production in that germinal centre formation in lymph nodes is not a marked feature, but there is a marked proliferation of immunoblasts and later of small lymphocytes in paracortical areas, with little evidence of mature plasma cell formation (Turk, 1967).

All these factors, taken together, suggest that following the introduction of antigen there are fundamental differences in the sequence of events leading to humoral antibody production and to delayed-type hypersensitivity.

The findings of Strober and Gowans (1965) indicate that peripheral sensitization of blood lymphocytes to antigens of renal allografts occurs; and Medawar (1965) has suggested that delayed-type hypersensitivity differs from humoral antibody production in that small lymphocytes become stimulated by antigen in the periphery by passing through the site of antigen administration, where they pick up antigen or immunogenic information, and migrate to draining lymph nodes where they transform to immunoblasts. The presence in such lymph nodes of large numbers of small lymphocytes having increased numbers of lysosomes may be interpreted as supporting evidence for this theory (Turk, 1967). However, macrophages are constantly present at sites of antigen injection and in the tissues' reaction to allografts, and it is conceivable that they localize antigen and subsequently pass immunogenic information to antigen-reactive lymphocytes in this location. In the present state of knowledge, there is little available evidence which either confirms or denies an essential role for macrophages in the induction of cellular immunity.

REFERENCES

Ackroyd, J. F. and Rook, A. (1963). Drug reactions. In *Clinical Aspects of Immunology* (eds. P. G. H. Gell and R. R. A. Coombs), pp. 448–96. Oxford: Blackwell.

Ada, G. L. and Lang, P. G. (1966). Antigens in tissues. II. State of antigen in lymph nodes of rats given isotopically-labelled flagellin, haemocyanin or serum albumin. *Immunology*, **10**, 431–43.

Ada, G. L., Nossal, G. J. V. and Pye, J. (1964). Antigens in immunity. III. Distribution of iodinated antigen following injection into rats via the hind footpads. *Aust. J. exp. Biol. med. Sci.* **42**, 295–310.

Ada, G. L. and Williams, J. M. (1966). Antigens in tissues. I. State of bacterial flagella in lymph nodes of rats injected with isotopically labelled flagella. *Immunology*, **10**, 417–29.

Adams, C. W. M. and Tuquan, N. A. (1961). Elastic degeneration as source of lipids in the early lesion of atherosclerosis. *J. Path. Bact.* **82**, 131–9.

Adler, F. L., Fishman, M. and Dray, S. (1966). Antibody formation initiated *in vitro*. III. Antibody formation and allotypic specificity directed by ribonucleic acid from peritoneal exudate cells. *J. Immun.* **97**, 554–8.

Al-Askari, S., David, J. R., Lawrence, H. S. and Thomas, L. (1965). *In vitro* studies on homograft sensitivity. *Nature, Lond.* **205**, 916–17.

Alexander, P., Delorme, E. J. and Hall, J. G. (1966). The effect of lymphoid cells from the lymph of specifically immunised sheep on the growth of primary sarcomata in rats. *Lancet*, **1**, 1186–9.

Alexander, P. and Fairley, G. H. (1967). Cellular resistance to tumours. *Br. med. Bull.* **23**, 86–92.

Allison, A. C. (1970). On the role of macrophages in some pathological processes. In *Mononuclear Phagocytes* (ed. R. Van Furth), pp. 422–40. Oxford: Blackwell.

Allison, A. C., Harington, J. S. and Birbeck, M. (1966). An examination of the cytotoxic effects of silica on macrophages. *J. exp. Med.* **124**, 141–54.

Amos, D. B. (1962). Host response to ascites tumours. In *Mechanism of Cell and Tissue Damage Produced by Immune Reactions* (2nd International Symposium on Immunopathology) (eds. P. Grabar and P. Miescher), pp. 210–22. Basel: Schwabe.

Amos, H. E., Gurner, B. W., Olds, R. J. and Coombs, R. R. A. (1967). Passive sensitization of tissue cells. II. Ability of cytophilic antibody to render the migration of guinea-pig peritoneal cells inhibitable by antigen. *Int. Archs Allergy appl. Immun.* **32**, 496–505.

André-Schwartz, J., Schwartz, R. S., Mirjl, L. and Beldotti, L. (1967). Neoplastic sequelae of allogenic disease. II. Electron-microscopic study of a neoplasm (Reticuloendotheliosis) in survivors of the Graft-vs-Host reaction. *Am. J. Pathol.* **50**, 707–44.

Antonini, F. M., Cappelli, G., Citi, S. and Serio, M. (1964). Alcune osservazione sui rapporti fra invecciamento e potere granulopessico del sistema reticulo-endotheliale. *Giorn. Geront.* **12**, 741–50.

Antonini, F. M., Weber, G. and Zampi, G. (1960). Biochemical and morphological studies on the hyperlipaemia, xanthomatosis, and endogenous experimental atherogenesis in rabbits, following a combined treatment with Tween 80 and cholesterol. In *Reticulo-endothelial Structure and Function* (ed. J. H. Heller), pp. 431–8. New York: Ronald.

Aschoff, L. (1924). Das Reticulo-endotheliale System. *Ergebn. inn. Med. Kinderheilk.* **26**, 1–118.

Asherson, G. L. and Zembala, M. (1970). The role of the macrophage in delayed-type hypersensitivity. In *Mononuclear Phagocytes* (ed. R. Van Furth), pp. 495–509. Oxford: Blackwell.

Ashworth, C. T., DiLuzio, N. R. and Riggi, S. J. (1963). A morphologic study of the effect of reticuloendothelial stimulation upon hepatic removal of minute particles from the blood of rats. *Exp. Mol. Pathol.* **2**, *Suppl.*, **1**, 83–103.

Askonas, B. A., Auzins, I. and Unanue, E. R. (1968). Role of macrophages in the immune response. *Bull. Soc. Chim. biol.* **50**, 1113–28.

Askonas, B. A. and Rhodes, J. M. (1965). Immunogenicity of antigen-containing ribonucleic acid preparation from macrophages. *Nature, Lond.* **205**, 470–4.

Auzins, I. and Rowley, D. (1962). On the question of the specificity of cellular immunity. *Aust. J. exp. Biol. med. Sci.* **40**, 283–92.

Axline, S. G. and Cohn, Z. A. (1970). *In vitro* induction of lysosomal enzymes by macrophages. *J. exp. Med.* **131**, 1239–60.

Bain, B., Lowenstein, L. and Maclean, L. D. (1965). The *in vitro* 'mixed leukocyte reaction' and initial studies on its application as a test for histoincompatibility. In *Histoincompatibility Testing* (National Research Council, Publ. No. 1229), pp. 121–9. Washington: National Academy of Sciences.

Balner, H. (1963). Identification of peritoneal macrophages in mouse radiation chimeras. *Transplantation*, **1**, 217–23.

Bangham, A. D. (1964). The adhesiveness of leucocytes with special reference to zeta potential. *Ann. N.Y. Acad. Sci.* **116**, 945–9.

Barnhart, M. I. and Cress, D. C. (1967). Plasma clearance of products of fibrinolysis. In *The Reticuloendothelial System and Atherosclerosis* (eds. N. R. DiLuzio and R. Paoletti), pp. 492–502. New York: Plenum.

Battaglia, S. (1962). Electronoptic studies in liver amyloid in mice (Ger.). *Beitr. path. Anat.* **126**, 300–20.

Benacerraf, B., Biozzi, G., Halpern, B. N. and Stiffel, C. (1957). Physiology of phagocytosis of particles by the R.E.S. In *Physiopathology of the Reticuloendothelial System* (eds. B. N. Halpern, B. Benacerraf and J. F. Delafresnaye), pp. 52–77. Oxford: Blackwell.

Benacerraf, B. and Sebestyen, M. M. (1957). Effect of bacterial endotoxins on the reticuloendothelial system. *Fed. Proc.* **16**, 860.

Benacerraf, B., Sebestyen, M. M. and Schlossman, S. (1959). A quantitative study of the kinetics of blood clearance of ^{32}P labelled *Escherichia coli* and Staphylococci by the reticulo-endothelial system. *J. exp. Med.* **110**, 27–48.

Bennett, B. (1965). Phagocytosis of mouse tumour cells *in vitro* by various homologous and heterologous cells. *J. Immun.* **95**, 80–6.

Bennett, B., Old, L. J. and Boyse, E. A. (1964). The phagocytosis of tumour cells *in vitro*. *Transplantation*, **2**, 183–202.

Bennett, W. E. and Cohn, Z. A. (1966). The isolation and selected properties of blood monocytes. *J. exp. Med.* **123**, 145–59.

Berg, J. W. (1956). Sinus histiocytosis. A fallacious measure of host resistance to cancer. *Cancer*, **9**, 935–9.

Bergenhem, B. and Fåhraeus, R. (1936). Uber spontane Hämolysinbildung im Blut, unter besonderer Berücksichtigung der Physiologie der Milz. *Z. exp. Med.* **97**, 555–87.

Berliner, D. L., Nabors, C. J. and Dougherty, T. F. (1964). The role of hepatic and adrenal reticuloendothelial cells in steroid biotransformation. *J. Reticuloendothel. Soc.* **1**, 1–17.

Berman, L. (1966). Lymphocytes and macrophages *in vitro*. Their activities in relation to functions of small lymphocytes. *Lab. Invest.* **15**, 1084–99.

Bernick, S. and Patek, P. R. (1961). Effect of cholesterol feeding on rat reticulo-endothelial system. *Archs Path.* **72**, 310–20.

Bessis, M. (1963). Quelques données cytologiques sur le rôle du système réticulo-endothélial dans l'erythropoïèse et l'érythroclasie. In *Rôle du Système Réticulo-endothélial dans l'immunité antibactérienne et antitumorale* (ed. B. N. Halpern), pp. 447–57. Paris: CNRS.

Bessis, M. and Breton-Gorius, J. (1962). Iron metabolism in the bone marrow as seen by electron microscopy: a critical review. *Blood*, **19**, 635–63.

Billingham, R. E., Defendi, V., Silvers, W. K. and Steinmuller, D. (1962). Quantitative studies on the induction of tolerance of skin homografts and on runt disease in neonatal rats. *J. natn. Cancer Inst.* **28**, 365–435.

Biozzi, G., Benacerraf, B. and Halpern, B. N. (1953). Quantitative study of the granulopectic activity of the reticuloendothelial system. II. A study of the kinetics of the granulopectic activity of the R.E.S. in relation to the dose of carbon injected. Relationship between the weight of organs and their activity. *Br. J. exp. Path.* **34**, 441–57.

Biozzi, G., Halpern, B. N., Benacerraf, B. and Stiffel, C. (1957). Phagocytic activity of the reticulo-endothelial system in experimental infections. In *Physiopathology of the Reticulo-endothelial System* (eds. B. N. Halpern, B. Benacerraf and J. F. Delafresnaye), pp. 204–24. Oxford: Blackwell.

Biozzi, G., Halpern, B. N. and Stiffel, C. (1963). Stimulation du système réticulo-endothélial (S.R.E.) par l'extrait microbien Wxb 3148 et résistance aux infections expérimentales. In *Rôle du Système Réticulo-endothélial dans l'immunité antibactérienne et antitumorale* (ed. B. N. Halpern), pp. 205–20. Paris: CNRS.

Biozzi, G., Howard, J. G., Halpern, B. N., Stiffel, C. and Mouton, D. (1960). The kinetics of blood clearance of isotopically labelled *Salmonella enteritidis* by the reticulo-endothelial system in mice. *Immunology*, **3**, 74–89.

Biozzi, G., Howard, J. G., Stiffel, C. and Mouton, D. (1964). The effect of splenectomy on the severity of graft-versus-host disease in adult mice. *J. Reticuloendothel. Soc.* **1**, 18–28.

Biozzi, G. and Stiffel, C. (1963). Étude du rôle des opsonines dans le phénomène de phagocytose des microorganismes et des colloides par les cellules du système réticulo-endothélial. In *Rôle du Système Réticulo-endothélial dans l'immunité antibactérienne et antitumorale* (ed. B. N. Halpern), pp. 262–86. Paris: CNRS.

Biozzi, G., Stiffel, C., Halpern, B. N. and Mouton, D. (1960). Recherches sur le mécanisme de l'immunité non specifique produite par les mycobactéries. *Revue fr. Etud. clin. biol.* **5**, 876–90.

Black, M. M. and Speer, F. D. (1958). Sinus histiocytosis of lymph nodes in cancer. *Surgery Gynec. Obstet.* **106**, 163–75.

Black, M. M. and Speer, F. D. (1960). Lymph node reactivity in cancer patients. *Surgery Gynec. Obstet.* **110**, 477–487.

Blanden, R. V., Lefford, M. J. and Mackaness, G. B. (1969). The host response to Calmette-Guérin Bacillus infection in mice. *J. exp. Med.* **129**, 1079–101.

Bloom, B. R. and Bennett, B. (1966). Mechanism of a reaction *in vitro* associated with delayed-type hypersensitivity. *Science, N.Y.* **153**, 80–2.

Bloom, W. (1927). Transformation of lymphocytes of thoracic duct into polyblasts (macrophages) in tissue culture. *Proc. Soc. exp. Biol. Med.* **24**, 567–9.

Boak, J. L., Christie, G. H., Ford, W. L. and Howard, J. G. (1968). Pathways in the development of liver macrophages; alternative precursors contained in populations of lymphocytes and bone marrow cells. *Proc. R. Soc. B.* **169**, 307–27.

Böhme, D. H. (1960). The reticuloendothelial system and nonspecific resistance. *Ann. N.Y. Acad. Sci.* **88**, 172–83.

Böhme, D. H. and Bouvier, C. A. (1960). The influence of killed *Mycobacterium tuberculosis* and its constituents upon the reticulo-endothelial system, internal organs, and percentage distribution of leukocytes in normal albino mice. In *Reticulo-endothelial Structure and Function* (ed. J. H. Heller), pp. 165–74. New York: Ronald.

Bolis, L., Kessel, R. W. I. and Petti, G. (1967). The action of some natural substances on the RES. In *The Reticuloendothelial System and Atherosclerosis* (eds. N. R. DiLuzio and R. Paoletti), pp. 197–202. New York: Plenum.

Bonnin, J. A. and Schwartz, L. (1954). The combined studies of agglutination, hemolysis, and erythrophagocytosis with special reference to acquired hemolytic anaemia. *Blood*, **9**, 773–88.

Boyden, S. V. (1962). The chemotactic effect of mixtures of antibody and antigen on polymorphonuclear leucocytes. *J. exp. Med.* **115**, 453–66.

Boyden, S. V. (1964). Cytophilic antibody in guinea-pigs with delayed-type hypersensitivity. *Immunology*, **7**, 474–83.

Boyden, S. V. (1966). Natural antibodies and the immune response. *Adv. Immunol.* **5**, 1–28.

Boyden, S. V., North, R. J. and Faulkner, S. M. (1965). Complement and the activity of phagocytes. In *Complement* (Ciba Foundation Symposium) (eds. G. W. Wolstenholme and J. Knight), pp. 190–213. London: Churchill.

Brady, R. O., Kaufer, J. N., Mock, M. B. and Fredrickson, D. S. (1966). The metabolism of sphingomyelin. II. Evidence of an enzymatic deficiency in Niemann–Pick disease. *Proc. natn. Acad. Sci. U.S.A.* **55**, 366–9.

Brent, L. and Medawar, P. B. (1967). Cellular immunity and the homograft reaction. *Br. med. Bull.* **23**, 55–60.

Brumfitt, W. and Glynn, A. A. (1961). Intracellular killing of *Micrococcus lysodeikticus* by macrophages and polymorphonuclear leucocytes. A comparative study. *Br. J. exp. Path.* **42**, 408–23.

Burkel, W. E. and Low, F. N. (1965). The fine structure of rat liver sinusoids, space of Dissé and associated tissue spaces. *Am. J. Anat.* **118**, 769–84.

Byers, S. O. (1960). Lipids and the reticuloendothelial system. *Ann. N.Y. Acad. Sci.* **88**, 240–3.

Byers, S. O., Mist-St George, S. and Friedman, M. (1957). Hepatic reticulo-endothelial cells as participants in the normal disposition of exogenous cholesterol in the rat. In *Physiopathology of the Reticulo-endothelial System* (eds. B. N. Halpern, B. Benacerraf and J. F. Delafresnaye), pp. 128–46. Oxford: Blackwell.

Carr, I. (1970). The fine structure of the mammalian lymphoreticular system. *Int. Rev. Cytol.* **27**, 283–348.

Carr, I., Clarke, J. A. and Salsbury, A. J. (1969). The surface structure of mouse peritoneal cells – a study with the scanning electron microscope. *J. Microscopy*, **89**, 105–11.

Carr, I. and Williams, M. A. (1967). The cellular basis of RE stimulation: the effects on peritoneal cells of stimulation with glyceryl trioleate, studied by EM and autoradiography. In *The Reticuloendothelial System and Atherosclerosis* (eds. N. R. DiLuzio and R. Paoletti), pp. 98–107. New York: Plenum.

Carter, R. L. and Gershon, R. K. (1966). Studies on homotransplantable lymphomas in hamsters. I. Histological responses in lymphoid tissues and their relationship to metastasis. *Am. J. Path.* **49**, 637–50.

Carter, R. L. and Gershon, R. K. (1967). Studies on homotransplantable lymphomas in hamsters. II. Further observations on the specificity of the histological responses in lymphoid tissues and their relationship to metastasis. *Am. J. Path.* **50**, 203–17.

Chare, M. J. B. and Boak, J. L. (1970). Effect on antilymphocyte serum on macrophage activity. *Clin. exp. Immunol.* **6**, 655–9.

Cohen, A. S. (1965). The constitution and genesis of amyloid. *Int. Rev. exp. Pathol.* **4**, 159–243.

Cohen, A. S., Gross, E. S. and Shirahama, T. (1964). Production of amyloid by tissue explants. *Arthritis Rheum.* **7**, 301.

Cohn, Z. A. (1962). Influence of polymorphonuclear leukocytes and macrophages on the immunogenicity of *Escherichia coli. Nature, Lond.* **196**, 1066–8.

Cohn, Z. A. (1963). The fate of bacteria within phagocytic cells. I. The degradation of isotopically labeled bacteria by polymorphonuclear leukocytes and macrophages. *J. exp. Med.* **117**, 27–42.

Cohn, Z. A. (1964). The fate of bacteria within phagocytic cells. III. Destruction of an *Escherichia coli* agglutinogen within polymorphonuclear leukocytes and macrophages. *J. exp. Med.* **120**, 869–83.

Cohn, Z. A. (1966). The regulation of pinocytosis in mouse macrophages. I. Metabolic requirements as defined by the use of inhibitors. *J. exp. Med.* **124**, 557–71.

Cohn, Z. A. (1968). The structure and function of monocytes and macrophages. *Adv. Immunol.* **9**, 163–214.

Cohn, Z. A., Fedorko, M. E. and Hirsch, J. G. (1966 a). The *in vitro* differentiation of mononuclear phagocytes. V. The formation of macrophage lysosomes. *J. exp. Med.* **123**, 757–66.

Cohn, Z. A., Hirsch, J. G. and Fedorko, M. E. (1966 b). The *in vitro* differentiation of mononuclear phagocytes. IV. The ultrastructure of macrophage differentiation in the peritoneal cavity and in culture. *J. exp. Med.* **123**, 747–56.

Cohn, Z. A. and Parks, E. (1967 a). The regulation of pinocytosis in mouse macrophages. II. Factors inducing vesicle formation. *J. exp. Med.* **125**, 213–30.

Cohn, Z. A. and Parks, E. (1967 b). The regulation of pinocytosis in mouse macrophages. III. The induction of vesicle formation by nucleosides and nucleotides. *J. exp. Med.* **125**, 457–66.

Cohn, Z. A. and Parks, E. (1967 c). The regulation of pinocytosis in mouse macrophages. IV. The immunological induction of pinocytotic vesicles, secondary lysosomes and hydrolytic enzymes. *J. exp. Med.* **125**, 1091–104.

Cohn, Z. A. and Wiener, E. (1963). The particulate hydrolases of macrophages. I. Comparative enzymology, isolation and properties. *J. exp. Med.* **118**, 991–1008.

Cole, L. J. and Garver, R. M. (1961). Homograft-reactive large mononuclear leukocytes in peripheral blood and peritoneal exudates. *Am. J. Physiol.* **200**, 147–51.

Coons, A. H., Leduc, E. H. and Kaplan, M. H. (1951). Localization of antigen in tissue cells. VI. The fate of injected foreign proteins in the mouse. *J. exp. Med.* **93**, 173–88.

Cooper, G. N. (1964). Functional modification of reticuloendothelial cells by simple triglycerides. *J. Reticuloendothel. Soc.* **1**, 50–67.

Cooper, G. N. and Howard, J. G. (1961). An effect of the graft-versus-host reaction on resistance to experimental bacteraemia. *Br. J. exp. Path.* **42**, 558–63.

Cooper, M. D., Peterson, R. D. A., South, M. A. and Good, R. A. (1966). The functions of the thymus system and the bursa system in the chicken. *J. exp. Med.* **123**, 75–102.

Crabbé, J. (1956). Enhancing action of small doses of cortisone on macrophage phagocytosis of staphylococci in rabbits. *Acta endocr., Copnh.* **21**, 41–6.

Cudkowicz, G., Upton, A. C., Shearer, G. M. and Hughes, W. L. (1964). Lymphocyte content and proliferative capacity of serially transplanted mouse bone marrow. *Nature, Lond.* **201**, 165–7.

Curran, R. C. (1967). Recent developments in inflammation and repair. In *Modern Trends in Pathology*. Vol. II (ed. T. Crawford), pp. 40–101. London: Butterworths.

Cutler, S. J., Black, M. M., Friedell, G. H., Vidone, R. A. and Goldeneberg, I. S. (1966). Prognostic factors in cancer of the female breast. II. Reproducibility of histopathologic classification. *Cancer,* **19**, 75–82.

Dacie, J. V. and Mollison, P. L. (1943). Survival of normal erythroctes after transfusion to patients with familial haemolytic anaemia. *Lancet,* **1**, 550–2.

Dannenberg, A. M. and Walter, P. C. (1961). A histochemical study of phagocytic and enzymatic functions of rabbit mononuclear (MN) and polymorphonuclear (PMN) exudate cells and alveolar macrophages (AM). *Fed. Proc.* **20**, 264.

Dannenberg, A. M., Walter, P. C. and Kapral, F. A. (1963). A histochemical study of phagocytic and enzymatic functions of rabbit mononuclear and polymorphonuclear exudate cells and alveolar macrophages. II. The effect of particle ingestion on enzymatic activity; two phases of *in vitro* activation. *J. Immun.* **90**, 448–65.

David, J. R. (1966). Delayed hypersensitivity *in vitro*: its mediation by cell-free substances formed by lymphoid cell–antigen interaction. *Proc. natn. Acad. Sci. U.S.A.* **56**, 72–7.

David, J. R. (1970). The role of macrophages in delayed hypersensitivity reactions. In *Mononuclear Phagocytes* (ed. R. Van Furth), pp. 486–93. Oxford: Blackwell.

David, J. R., Al-Askari, S., Lawrence, H. S. and Thomas, L. (1964). Delayed hypersensitivity *in vitro*. I. The specificity of inhibition of cell migration by antigens. *J. Immun.* **93**, 264–73.

David, J. R., Lawrence, H. S. and Thomas, L. (1964). The *in vitro* desensitization of sensitive cells by trypsin. *J. exp. Med.* **120**, 1189–1200.

Day, A. J. (1964). The macrophage system, lipid metabolism and atherosclerosis. *J. Atheroscler. Res.* **4**, 117–30.

de Duve, C. and Wattiaux, R. (1966). Functions of lysosomes. *A. Rev. Physiol.* **28**, 435–92.

Delorme, E. J. and Alexander, P. (1964). Treatment of primary fibrosarcoma in the rat with immune lymphocytes. *Lancet,* **2**, 117–21.

Del Rio Hortega, P. and dé Asua, F. (1924). Sobre las cellulas del reticulo esplenico y sur relaciones con el endothelio sinusal. *Boln. R. Soc. esp. Hist. nat.* **11**, 7.

Di Carlo, F. J., Haynes, L. J. and Phillips, G. E. (1963). Effect of *Mycobacterium phlei* upon reticuloendothelial system of mice of different ages. *Proc. Soc. exp. Biol. Med.* **112**, 651–5.

Diengdoh, J. V. and Turk, J. L. (1965). Immunological significance of lysosomes within lymphocytes *in vivo*. *Nature, Lond.* **207**, 1405–6.

DiLuzio, N. R. (1959). Lipid composition of Kupffer cells. *Am. J. Physiol.* **196**, 884–6.

DiLuzio, N. R. (1960). Reticuloendothelial involvement in lipid metabolism. *Ann. N.Y. Acad. Sci.* **88**, 244–51.

DiLuzio, N. R. and Blickens, D. A. (1966). Influence of intravenously administered lipids on reticulo-endothelial function. *J. Reticuloendothel. Soc.* **3**, 250–70.

DiLuzio, N. R. and Riggi, S. J. (1967). Participation of hepatic parenchymal and Kupffer cells in chylomicron and cholesterol metabolism. In *The Reticuloendothelial System and Atherosclerosis* (eds. N. R. DiLuzio and R. Paoletti), pp. 382–403. New York: Plenum.

Dobson, E. L. (1957). Factors controlling phagocytosis. In *Physiopathology of the Reticulo-endothelial System* (eds. B. N. Halpern, B. Benacerraf and J. F. Delafresnaye), pp. 80–113. Oxford: Blackwell.

Dobson, E. L., Kelly, L. S. and Finney, C. R. (1967). Kinetics of phagocytosis of repeated injections of colloidal carbon: blockade, a latent period or stimulation? A question of timing and dose. In *The Reticuloendothelial System and Atherosclerosis* (eds. N. R. DiLuzio and R. Paoletti), pp. 63–73. New York: Plenum.

Donaldson, D. H., Marais, S., Gyi, K. K. and Perkins, E. H. (1956). The influence of immunization and total body X-irradiation on intracellular digestion by peritoneal phagocytes. *J. Immun.* **76**, 192–9.

Dougherty, T. F. (1961). Role of steroids in regulation of inflammation. In *Inflammation and Diseases of Connective Tissues* (eds. L. C. Mills and J. H. Moyer), pp. 449–60. Philadelphia: Saunders.

Dresser, D. W. (1961). Effectiveness of lipid and lipidophilic substances as adjuvants. *Nature, Lond.* **191**, 1169–71.

Dresser, D. W. (1962). Specific inhibition of antibody production. II. Paralysis induced in adult mice by small quantities of protein antigen. *Immunology*, **5**, 378–88.

Dumonde, D. C. (1967). The role of the macrophage in delayed hypersensitivity. *Br. med. Bull.* **23**, 9–14.

Dumonde, D. C. and Medawar, P. B. (1967). Unpublished work quoted by Brent and Medawar (1967).

Dunnigan, M. G. (1964). The distribution of phospholipid within macrophages in human atheromatous plaques. *J. Atheroscler. Res.* **4**, 144–50.

Dupuy, J. M., Perey, D. U. E. and Good, R. A. (1969). Passive transfer with plasma, of delayed allergy in guinea-pigs. *Lancet*, **1**, 551–3.

Ebert, R. H. and Florey, H. W. (1939). The extravascular development of the monocyte observed *in vivo*. *Br. J. exp. Path.* **20**, 342–56.

Edwards, V. D. and Simon, G. T. (1970). Ultrastructural aspects of red cell destruction in the normal rat spleen. *J. Ultrastruct. Res.* **33**, 187–201.

Ehrenreich, B. A. and Cohn, Z. A. (1967). The uptake and digestion of iodinated human serum albumin by macrophages *in vitro*. *J. exp. Med.* **126**, 941–58.

Ehrenstein, G. von and Lockner, D. (1958). Sites of the physiological breakdown of the red blood corpuscles. *Nature, Lond.* **181**, 911.

Ehrich, W., Drabkin, D. L. and Forman, C. (1949). Nucleic acids and the formation of antibodies by plasma cells. *J. exp. Med.* **90**, 157–67.

Ehrlich, P. (1891). Farbenanalatische untersuchungen zur Histologie und Klinik des Blutes. Berlin: Hirschwald.

Eyring, H. and Dougherty, T. F. (1955). Molecular mechanism in inflammation and stress. *Am. Scient.* **43**, 457–67.

Fagraeus, A. (1948). Antibody production in relation to the development of plasma cells. *Acta med. scand.* **130** (Suppl. 204), 5–122.

Fauve, R. M. and Delaunay, A. (1966). Résistance cellulaire à l'infection bactérienne. III. Modifications de la résistance de souris N.C.S. à l'infection par *Listeria monocytogenes* après injection d'endotoxine. *Annls. Inst. Pasteur, Paris*, **110**, 95–105.

Filkins, J. P. and DiLuzio, N. R. (1966). Mechanism of gelatin inhibition of reticuloendothelial function. *Proc. Soc. exp. Biol. Med.* **122**, 177–80.

Filkins, J. P., Saba, T. M. and DiLuzio, N. R. (1966). The role of opsonins in the *in vitro* phagocytosis of colloidal gold by hepatic macrophages. *Fed. Proc.* **25**, 479.

Fischer, H., Ax, W., Freund-Mölbert, E., Holub, M., Krüsmann, W. F. and Matthes, M. L. (1970). Studies on phagocytic cells of the omentum. In *Mononuclear Phagocytes* (ed. R. Van Furth), pp. 528–47. Oxford: Blackwell.

Fishman, M. and Adler, F. L. (1963). Antibody formation *in vitro*. In *Immuno-pathology* (3rd International Symposium) (eds. P. Grabar and P. A. Miescher), pp. 79–88. Basel: Schwabe.

Fishman, M. and Adler, F. L. (1970). Heterogeneity of macrophage functions in relation to the immune response. In *Mononuclear Phagocytes* (ed. R. Van Furth), pp. 581–92. Oxford: Blackwell.

Fishman, M., van Rood, J. J. and Adler, F. L. (1965). The initiation of antibody formation by ribonucleic acid from specifically stimulated macrophages. In *Molecular and Cellular Basis of Antibody Formation* (ed. J. Sterzl), pp. 491–8. Prague: Czechoslovak Academy of Sciences.

Foley, E. J. (1953). Attempts to induce immunity against mammary adenocarcinoma in inbred mice. *Cancer Res.* **13**, 578–80.

Forbes, I. J. (1965). Mitosis in mouse peritoneal macrophages. *J. Immun.* **96**, 734–43.

Ford, W. F., Gowans, J. L. and McCullagh, P. J. (1966). The origin and function of lymphocytes. In *The Thymus: Experimental and Clinical Studies* (Ciba Foundation Symposium) (eds. G. E. W. Wolstenholme and R. Porter), pp. 58–79. Boston: Little Brown.

Fox, M. (1962). Cytological estimation of proliferating donor cells during graft-versus-host disease in F_1 hybrid mice injected with parental spleen cells. *Immunology*, **5**, 489–95.

Fred, R. K. and Shore, M. L. (1967). Application of a mathematical model to the study of RES phagocytosis in mice. In *The Reticuloendothelial System and Atherosclerosis* (eds. N. R. DiLuzio and R. Paoletti), pp. 1–17. New York: Plenum.

Fredrickson, D. S., Levy, R. I. and Lees, R. S. (1967). Fat transport in lipoproteins – an integrated approach to mechanisms and disorders. *New Engl. J. Med.* **276**, 34–44, 94–103, 148–56, 215–24, 273–81.

Freedman, H. H. (1960). Reticuloendothelial system and passive transfer of endo-toxin tolerance. *Ann. N.Y. Acad. Sci.* **88**, 99–106.

Frei, P. C., Benacerraf, B. and Thorbecke, G. J. (1965). Phagocytosis of the antigen. A crucial step in the induction of the primary response. *Proc. natn. Acad. Sci., U.S.A.* **53**, 20–3.

French, J. E. (1966). Atherosclerosis in relation to the structure and function of the arterial intima, with special reference to the endothelium. *Int. Rev. exp. Pathol.* **5**, 253–353.

French, J. E. and Morris, B. (1957). The removal of ^{14}C-labelled chylomicron fat from the circulation in rats. *J. Physiol., Lond.* **138**, 326–39.

Friedman, H. P., Stavitsky, A. B. and Solomon, J. M. (1965). Interaction *in vitro* of antibodies to phage T2: antigens in the RNA extract employed. *Science, N.Y.* **149**, 1106–7.

Friedman, M., Byers, S. O. and Rosenman, R. H. (1954). Demonstration of the essential role of the hepatic reticulo-endothelial cell (Kupffer cell) in the normal disposition of exogenously derived cholesterol. *Am. J. Physiol.* **177**, 77–83.

Gabrielli, E. R. and Snell, F. M. (1965). Reflection of reticulo-endothelial function in studies of blood clearance kinetics. *J. Reticuloendothel. Soc.* **2**, 141–57.

Garvin, J. E. (1961). Factors affecting the adhesiveness of human leucocytes and platelets *in vitro*. *J. exp. Med.* **114**, 51–73.

George, M. and Vaughan, J. H. (1962). *In vitro* cell migration as a model for delayed hypersensitivity. *Proc. Soc. exp. Biol. Med.* **111**, 514–21.

Gillman, T. and Wright, L. J. (1966). Auto-radiographic evidence suggesting *in vivo* transformation of some blood mononuclears in repair and fibrosis. *Nature, Lond.* **209**, 1086–90.

Gilman, R. and Trowell, O. A. (1965). The effect of radiation on the activity of reticuloendothelial cells in organ cultures of lymph nodes and thymus. *Int. J. Radiat. Biol.* **9**, 313–22.

Glynn, A. A., Brumfitt, W. and Salton, M. R. J. (1966). The specific activity and specific inhibition of intracellular lysozyme. *Br. J. exp. Path.* **47**, 331–6.

Goodman, J. W. (1964). On the origin of peritoneal fluid cells. *Blood,* **23**, 18–26.

Gordon, L. E., Cooper, D. B. and Miller, C. P. (1955). Clearance of bacteria from the blood of irradiated rabbits. *Proc. Soc. exp. Biol. Med.* **89**, 577–9.

Gorer, P. A. (1956). Some recent work on tumour immunity. *Adv. Cancer Res.* **4**, 149–80.

Gorer, P. A. and Amos, D. B. (1956). Passive immunity in mice against C57BL leukosis E.L.4 by means of iso-immune serum. *Cancer Res.* **16**, 338–43.

Gorer, P. A., Loutit, J. F. and Micklem, H. S. (1961). Proposed revisions of 'transplantation'. *Nature, Lond.* **189**, 1024–5.

Gorstein, F. and Benacerraf, B. (1960). Hyperactivity of the reticuloendothelial system and experimental anaemia in mice. *Am. J. Path.* **37**, 569–82.

Gough, J., Elves, M. W. and Israëls, M. C. G. (1965). The formation of macrophages from lymphocytes *in vitro. Expl Cell Res.* **38**, 476–82.

Gowans, J. L. (1962). The fate of parental strain small lymphocytes in F_1 hybrid rats. *Ann. N.Y. Acad. Sci.* **99**, 432–55.

Gowans, J. L. (1966). Life-span, recirculation, and transformation of lymphocytes. *Int. Rev. exp. Pathol.* **5**, 1–24.

Grampa, G. (1967). Reticuloendothelial system stimulation by estrogens and thorium dioxide retention in rat liver. In *The Reticuloendothelial System and Atherosclerosis* (eds. N. R. DiLuzio and R. Paoletti), pp. 214–20. New York: Plenum.

Granger, G. A. and Weiser, R. S. (1964). Homograft target cells. Specific destruction *in vitro* by contact interaction with immune macrophages. *Science, N.Y.* **145**, 1427–9.

Granger, G. A. and Weiser, R. S. (1966). Homograft target cells. Contact destruction *in vitro* by immune macrophages. *Science, N.Y.* **151**, 97–9.

Hall, J. G. (1967). Quantitative aspects of the recirculation of lymphocytes: an analysis of data from experiments on sheep. *Q. Jl exp. Physiol.* **52**, 76–85.

Hall, J. G. and Morris, B. (1963). The lymph-borne cells of the immune response. *Q. Jl exp. Physiol.* **48**, 235–47.

Hall, J. W. and Furth, J. (1938). Cultural studies on the relationship of lymphocytes to monocytes and fibroblasts. *Archs Path.* **25**, 46–59.

Halpern, B. N., Biozzi, G. and Stiffel, C. (1963). Action de l'extrait microbien Wxb 3148 sur l'évolution des tumeurs expérimentales. In *Rôle du Système Réticuloendothélial dans l'immunité antibactérienne et antitumorale* (ed. B. N. Halpern), pp. 221–36. Paris: CNRS.

Han, S. S. (1961). The ultrastructure of the mesenteric lymph nodes of the rat. *Am. J. Anat.* **109**, 183–97.

Hand, R. A. and Chandler, A. B. (1962). Atherosclerotic metamorphosis of autologous pulmonary thromboemboli in the rabbit. *Am. J. Path.* **40**, 469–86.

Hanna, M. G. and Szakel, A. K. (1968). Localization of ^{125}I-labeled antigen in germinal centers of mouse spleen: histologic and ultrastructural autoradiographic studies of the secondary immune response. *J. Immun.* **101**, 949–62.

Harris, G. (1966). Ribonucleic acid synthesis in macrophages in relation to the secondary immune response *in vitro. Nature, Lond.* **211**, 154–5.

Harris, H. (1953). Chemotaxis of monocytes. *Br. J. exp. Path.* **34**, 276–9.

Havel, R. J. and Fredrickson, D. S. (1956). Metabolism of chylomicra; the removal of palmitic acid-1-C^{14}-labeled chylomicra from dog plasma. *J. clin. Invest.* **35**, 1025–32.

Hechter, O., Zaffaroni, A., Jacobsen, R. P., Levy, H., Jeanloz, R. W., Schenker, V. and Pincus, G. (1951). The nature and biosynthesis of the adrenal secretory products. *Recent Prog. Horm. Res.* **6**, 215–46.

Heise, E. R. and Weiser, R. S. (1967). *In vitro* studies on the mechanism of antigen-mediated inhibition of macrophage migration. *J. Reticuloendothel. Soc.* **4**, 450.

Heller, J. H., Meier, R. M., Zucker, R. and Mast, G. W. (1957). The effect of natural and synthetic estrogens on reticuloendothelial system function. *Endocrinology*, **61**, 235–41.

Helmy, I. D. and Nicol, T. (1951). The effect of oestrogenic hormones on the mesothelial cells of the peritoneum. *J. Obstet. Gynaec. Br. Commonw.* **58**, 588–90.

Hirsch, J. G., Fedorko, M. E. and Dwyer, C. M. (1966). The ultrastructure of epithelioid and giant cells in positive Kveim test sites and sarcoid granulomata. In *La Sarcoïdose* (4th International Conference on Sarcoidosis) (eds. J. Turiaf and J. Chabot), pp. 59–70. Paris: Masson.

Hollander, J. L., Fudenberg, H. H., Rawson, A. J., Abelson, N. M. and Torralba, T. P. (1966). Further studies on the pathogenesis of rheumatoid joint inflammation. *Arthritis Rheum.* **9**, 675–81.

Hollander, J. L., McCarty, D. J., Astorga, G. and Castro-Murillo, E. (1965). Studies on the pathogenesis of rheumatoid joint inflammation. I. The 'R.A. Cell' and a working hypothesis. *Ann. intern. Med.* **62**, 271–80.

Holter, H. (1959). Pinocytosis. *Int. Rev. Cytol.* **8**, 481–504.

Howard, J. G. (1961 *a*). Changes in the activity of the reticulo-endothelial system following the injection of parental spleen cells into F_1 hybrid mice. *Br. J. exp. Path.* **42**, 72–82.

Howard, J. G. (1961 *b*). Increased sensitivity to bacterial endotoxin of F_1 hybrid mice undergoing graft-versus-host reaction. *Nature, Lond.* **190**, 1122.

Howard, J. G. (1963). A study of reticulo-endothelial function during graft-versus-host reaction in the adult mouse. In *Rôle du Système Réticulo-endothélial dans l'immunité antibactérienne et antitumorale* (ed. B. N. Halpern), pp. 341–59. Paris: CNRS.

Howard, J. G. (1970). The origin and immunological significance of Kupffer cells. In *Mononuclear Phagocytes* (ed. R. van Furth), pp. 178–99. Oxford: Blackwell.

Howard, J. G., Boak, J. L., Christie, G. H. and Kinsky, R. G. (1969). Peritoneal and alveolar macrophages derived from lymphocyte populations during graft-versus-host reactions. *Br. J. exp. Path.* **50**, 448–55.

Howard, J. G., Christie, G. H., Boak, J. L. and Evans-Anfom, E. (1965). Evidence for the conversion of lymphocytes into liver macrophages during graft-versus-host reactions. *Colloques. int. Cent. natn. Rech. scient.* **147**, 95–102.

Howard, J. G. and Wardlaw, A. C. (1958). The opsonic effect of normal serum on the uptake of bacteria by the reticulo-endothelial system. *Immunology*, **1**, 338–52.

Howard, J. G. and Woodruff, M. F. A. (1961). Effect of the graft-versus-host reaction on the immunological responsiveness of the mouse. *Proc. R. Soc. B.* **154**, 532–9.

Hoy, W. E. and Nelson, D. S. (1967). Quoted by Nelson (1969).

Humphrey, J. H. and Frank, M. M. (1967). The localization of non-microbial antigens in the draining lymph nodes of tolerant, normal and primed rabbits. *Immunology*, **13**, 87–100.

Hutchin, P., Amos, D. B. and Prioleau, W. H. (1967). Interactions of humoral antibodies and immune lymphocytes. *Transplantation*, **5**, 68–75.

Hyman, C. and Paldino, R. L. (1960). Possible role of the reticuloendothelial system in protein transport. *Ann. N.Y. Acad. Sci.* **88**, 232–9.

Ilio, M. and Wagner, H. N. (1963). Studies of the reticuloendothelial system (RES). I. Measurement of the phagocytic capacity of the RES in man and dog. *J. clin. Invest.* **42**, 417–26.

Jacoby, F. (1965). Macrophages. In *Cells and Tissues in Culture*. Vol. II (ed. E. N. Willmer), pp. 1–93. New York & London: Academic Press.

Jaffe, H. L. and Lichtenstein, L. (1944). Eosinophilic granuloma of bone. *Archs Path.* **37**, 99–118.

Jaffé, R. H. (1938). The reticulo-endothelial system. In *Handbook of Haematology* (ed. H. Downey), pp. 974–1271. New York: Harper & Row (Hoeber).

Jenkin, C. R. (1963). The effect of opsonins on the intracellular survival of bacteria. *Br. J. exp. Path.* **44**, 47–57.

Jenkin, C. R. and Benacerraf, B. (1960). *In vitro* studies on the interaction between mouse peritoneal macrophages and strains of *Salmonella* and *Escherichia coli*. *J. exp. Med.* **112**, 403–17.

Jenkin, C. R. and Rowley, D. (1961). The role of opsonins in the clearance of living and inert particles by cells of the reticulo-endothelial system. *J. exp. Med.* **114**, 363–74.

Jenkin, C. R. and Rowley, D. (1963). *Salmonella typhimurium*, a parasite of the reticulo-endothelial system. In *Rôle du Système Réticulo-endothélial dans l'immunité antibactérienne et antitumorale* (ed. B. N. Halpern), pp. 291–318. Paris: CNRS.

Jerne, N. K., Nordin, A. A. and Henry, C. (1963). The agar plaque technique for recognizing antibody-producing cells. In *Cell-bound Antibodies* (eds. B. Amos and H. Koprowski), pp. 109–16. Philadelphia: Wistar Institute.

Jeter, W. S., McKee, A. P. and Mason, R. L. (1961). Inhibition of immune phagocytosis of *Diplococcus pneumoniae* by human neutrophiles with antibody against complement. *J. Immun.* **86**, 386–91.

Johnson, T. M. and Garvin, J. E. (1959). Separation of lymphocytes in human blood by means of glass wool. *Proc. Soc. exp. Biol. Med.* **102**, 333–5.

Jones, R., Thomas, W. A. and Scott, R. F. (1962). Electron microscopy study of chyle from rats fed butter or corn oil. *Exp. Mol. Pathol.* **1**, 65–83.

Journey, L. J. and Amos, D. B. (1962). An electron microscope study of histiocyte responses to ascites tumour homografts. *Cancer Res.* **22**, 998–1001.

Karnovsky, M. L. (1962). Metabolic basis of phagocytic activity. *Physiol. Rev.* **42**, 143–68.

Karnovsky, M. L., Shafer, A. W., Cagan, R. H., Graham, R. C., Karnovsky, M. J., Glass, E. A. and Saito, K. (1966). Membrane function and metabolism in phagocytic cells. *Trans N.Y. Acad. Sci.* **28**, 778–87.

Karnovsky, M. L. and Wallach, D. F. H. (1963). Metabolic responses of mammalian cells to solid particles. In *Rôle du Système Réticulo-endothélial dans l'immunité antibactérienne et antitumorale* (ed. B. N. Halpern), pp. 147–57. Paris: CNRS.

Karthigasu, K. and Jenkin, C. R. (1963). The functional development of the reticulo-endothelial system of the chick embryo. *Immunology*, **6**, 255–63.

Keller, H. U. and Sorkin, E. (1967). Studies on chemotaxis. V. On the chemotactic effect of bacteria. *Int. Archs. Allergy appl. Immun.* **31**, 505–17.

Kelly, L. S., Brown, B. A. and Dobson, E. L. (1962). Cell division and phagocytic activity in liver reticulo-endothelial cells. *Proc. Soc. exp. Biol. Med.* **110**, 555–9.

Kelly, L. S., Dobson, E. L., Finney, C. R. and Hirsch, J. D. (1960). Proliferation of the RES in the liver. *Am. J. Physiol.* **198**, 1134–8.

Khoo, K. K. and Mackaness, G. B. (1964). Macrophage proliferation in relation to acquired cellular resistance. *Aust. J. exp. Biol. med. Sci.* **42**, 707–16.

Klein, G., Sjögren, H. O., Klein, E. and Hellström, K. E. (1960). Demonstration of resistance against methylocholanthrene-induced sarcomas in the primary autochthonous host. *Cancer Res.* **20**, 1561–72.

Kojima, M. and Imai, Y. (1961). The mechanism of phagocytosis of the reticulo-endothelial cells. *Proc. Jap. Soc. RES.* **1**, 61–71.

Kölsch, E. and Mitchison, N. A. (1968). The subcellular distribution of antigens. *J. exp. Med.* **128**, 1059–79.

Kosunen, T. U. and Flax, M. H. (1966). Experimental allergic thyroiditis in the guinea pig. IV. Autoradiographic studies of the evolution of the cellular infiltrate. *Lab. Invest.* **15**, 606–16.

Kosunen, T. U., Waksman, B. H., Flax, M. H. and Tihen, W. S. (1963). Radioauto-graphic study of cellular mechanisms in delayed hypersensitivity. I. Delayed reactions to tuberculin and purified proteins in the rat and guinea pig. *Immunology*, **6**, 276–90.

Kosunen, T. U., Waksman, B. H. and Samuelsson, I. K. (1963). Radioautographic study of cellular mechanisms in delayed hypersensitivity. II. Experimental allergic encephalomyelitis in the rat. *J. Neuropathol. exp. Neurol.* **22**, 367–80.

Kountz, S. L., Williams, M. A., Williams, P. L., Kapros, G. and Dempster, W. J. (1963). Mechanism of rejection of homotransplanted kidneys. *Nature, Lond.* **199**, 257–60.

Leake, E. S., Gonzalez-Ojeda, O. and Myrvik, Q. N. (1964). Enzymatic differences between normal alveolar macrophages and oil-induced peritoneal macrophages obtained from rabbits. *Expl Cell Res.* **33**, 553–61.

Leake, E. S. and Myrvik, Q. N. (1966). Digestive vacuole formation in alveolar macrophages after phagocytosis of *Mycobacterium smegmatis in vivo. J. Reticulo-endothel. Soc.* **3**, 83–100.

Leduc, E. H., Coons, A. H. and Connolly, M. H. (1955). Studies on antibody production. II. The primary and secondary response in the popliteal lymph nodes of the rabbit. *J. exp. Med.* **102**, 61–72.

Lee, L. and McCluskey, R. T. (1962). Immunohistochemical demonstration of the reticuloendothelial clearance of circulating fibrin aggregates. *J. exp. Med.* **116**, 611–18.

Lewis, W. H. (1931). Pinocytosis. *Bull. Johns Hopkins Hosp.* **49**, 17–26.

Lichtenstein, L. (1953). Histiocytosis X. *Archs Path.* **56**, 84–102.

Lirenman, D. S., Fish, A. J. and Good, R. A. (1967). Lack of a serum factor in preventing blockade of the reticuloendothelial system in the rabbit. *J. Reticulo-endothel. Soc.* **4**, 34–42.

Loewi, G., Temple, A., Nind, A. P. P. and Axelrad, M. (1969). A study of the effect of anti-macrophage sera. *Immunology*, **16**, 99–106.

Looke, E. and Rowley, D. (1962). The lack of correlation between sensitivity of bacteria to killing by macrophages or acidic conditions. *Aust. J. exp. Biol. med. Sci.* **40**, 315–20.

Lozzio, B. B., Machado, E. and Lew, V. (1966). Red cell destruction in rats in RES hyperfunction produced by *p*-dimethylaminobenzene. *J. Reticuloendothel. Soc.* **3**, 149–62.

Lurie, M. B. (1960). The reticuloendothelial system, cortisone and thyroid function. *Ann. N.Y. Acad. Sci.* **88**, 83–98.

Lurie, M. B., Zappasodi, P., Dannenberg, A. M. and Schwartz, I. B. (1951). Con-stitutional factors in resistance to infection: effect of cortisone on pathogenesis of tuberculosis. *Science, N.Y.* **113**, 234–7.

Mabry, D. S., Bass, J. A., Dodd, M. C., Wallace, J. H. and Wright, C. S. (1956). Opsonic factors in normal and immune sera in the differential phagocytosis of normal, trypsinized and virus treated human and rabbit erythrocytes by macrophages in tissue culture. *J. Immun.* **76**, 54–61.

McDevitt, H. O., Askonas, B. A., Humphrey, J. H., Schechter, I. and Sela, M. (1966). The localization of antigen in relation to specific antibody-producing cells. I. Use of a synthetic polypeptide [(T,G)–A–L] labelled with iodine–125. *Immunology*, **11**, 337–51.

McFarland, W. and Heilman, H. (1965). Lymphocyte foot appendage: its role in lymphocyte function and in immunological reactions. *Nature, Lond.* **205**, 887–8.

Mackaness, G. B. (1967). The relationship of delayed hypersensitivity to acquired cellular resistance. *Br. med. Bull.* **23**, 52–4.

Mackaness, G. B. (1969). The influence of immunologically committed lymphoid cells on macrophage activity *in vivo. J. exp. Med.* **129**, 973–92.

Mackaness, G. B. (1970*a*). Cellular immunity. In *Mononuclear Phagocytes* (ed. R. Van Furth), pp. 461–75. Oxford: Blackwell.

Mackaness, G. B. (1970*b*). The monocyte in delayed-type hypersensitivity. In *Mononuclear Phagocytes* (ed. R. Van Furth), pp. 478–84. Oxford: Blackwell.

Mackaness, G. B. and Blanden, R. V. (1966). Cellular Immunity. *Prog. Allergy*, **11**, 89–140.

Mackaness, G. B. and Hill, W. C. (1969). The effect of anti-lymphocyte globulin on cell-mediated resistance to infection. *J. exp. Med.* **129**, 993–1012.

Marshall, A. H. E. (1956). *An Outline of the Cytology and Pathology of the Reticular Tissue*. Edinburgh: Oliver & Boyd.

Marshall, A. H. E. and White, R. G. (1950). Reactions of the reticular tissue to antigen. *Br. J. exp. Path.* **31**, 157–74.

Martin, W. J. (1966). The cellular basis of immunological tolerance in newborn animals. *Aust. J. exp. Biol. med. Sci.* **44**, 605–8.

Maximow, A. A. (1932). The macrophages or histiocytes. In *Special Cytology*. Vol. II, (ed. E. V. Cowdrey), pp. 709–70. New York: Haffner.

Maximow, A. A. (1962). *A Textbook of Histology*. 8th ed. Philadelphia: Saunders.

Medawar, J. (1940). Observations on lymphocytes in tissue culture. *Br. J. exp. Path.* **21**, 205–11.

Medawar, P. B. (1965). Transplantation of tissues and organs. Introduction. *Br. med. Bull.* **21**, 97–9.

Metcalf, D. (1966). *The Thymus*, Berlin: Springer-Verlag.

Metchnikoff, E. (1905). *Immunity in Infective Diseases*. London: Cambridge University Press.

Miescher, P. (1957). The role of the reticulo-endothelial system in haematoclasia. In *Physiopathology of the Reticulo-endothelial System* (eds. B. N. Halpern, B. Benacerraf and J. F. Delafresnaye), pp. 147–71. Oxford: Blackwell.

Mikulska, Z. B., Smith, C. and Alexander, P. (1966). Evidence for an immunological reaction of the host against an actively growing primary tumour. *J. natn. Cancer Inst.* **36**, 29–35.

Miller, J. F. A. P. and Mitchell, G. F. (1968). Cell to cell interaction in the immune response. I. Hemolysin-forming cells in neonatally thymectomized mice reconstituted with thymus or thoracic duct lymphocytes. *J. exp. Med.* **128**, 801–20.

Miller, J. J. and Nossal, G. J. V. (1964). Antigens in immunity. VI. The phagocytic reticulum of lymph node follicles. *J. exp. Med.* **120**, 1075–85.

Mims, C. A. (1962). Experiments on the origin and fate of lymphocytes. *Br. J. exp. Path.* **43**, 639–49.

Mims, C. A. (1964). Aspects of the pathogenesis of virus diseases. *Bact. Rev.* **28**, 30–71.

Mitchell, G. F. and Miller, J. F. A. P. (1968). Cell to cell interaction in the immune response. II. The source of hemolysin-forming cells in irradiated mice given bone marrow and thymus or thoracic duct lymphocytes. *J. exp. Med.* **128**, 821–37.

Mitchell, J. and Nossal, G. J. V. (1966). Mechanisms of induction of immunological tolerance. I. Localization of tolerance-inducing antigen. *Aust. J. exp. Biol. med. Sci.* **44**, 211–23.

Mitchison, N. A. (1969). The immunogenic capacity of antigen taken up by peritoneal exudate cells. *Immunology*, **16**, 1–14.

Möller, G. and Möller, E. (1967). Humoral and cell-mediated effector mechanisms in tissue transplantation. *J. clin. Path.* **20**, 437–50.

Morris, B. and French, J. E. (1958). The uptake and metabolism of ^{14}C labelled chylomicron fat by the isolated perfused liver of the rat. *Q. Jl exp. Physiol.* **43**, 180–8.

Murray, I. M. (1963). The mechanism of blockade of the reticulo-endothelial system. *J. exp. Med.* **117**, 139–47.

Myrvik, Q. N., Leake, E. S. and Oshima, S. (1962). A study of macrophages and epithelioid-like cells from granulomatous (BCG-induced) lungs of rabbits. *J. Immun.* **89**, 745–51.

Nash, T., Allison, A. C. and Harington, J. S. (1967). Physico-chemical properties of silica in relation to its toxicity. *Nature, Lond.* **210**, 259–61.

Nelson, D. S. (1969). *Macrophages and Immunity.* Amsterdam & London: North-Holland.

Nelson, D. S. and Boyden, S. V. (1963). The loss of macrophages from peritoneal exudates following the injection of antigens into guinea-pigs with delayed-type hypersensitivity. *Immunology*, **6**, 264–75.

Nelson, D. S. and Boyden, S. V. (1967). Macrophage cytophilic antibodies and delayed hypersensitivity. *Br. med. Bull.* **23**, 15–20.

Nicol, T. (1935). The female reproductive system in the guinea-pig: fat production: influence of hormones. *Trans. R. Soc. Edinb.* **58**, 449–86.

Nicol, T. and Bilbey, D. L. J. (1960). The effect of various steroids on the phagocytic activity of the reticuloendothelial system. In *Reticuloendothelial Structure and Function* (ed. J. H. Heller), pp. 301–20. New York: Ronald.

Nicol, T., Bilbey, D. L. J., Charles, L. M., Cordingley, J. L. and Vernon-Roberts, B. (1964). Oestrogen: the natural stimulant of body defence. *J. Endocr.* **30**, 277–91.

Nicol, T. and Cordingley, J. L. (1967). Reticuloendothelial excretion via the bronchial tree. In *The Reticuloendothelial System and Atherosclerosis* (eds. N. R. DiLuzio and R. Paoletti), pp. 58–62. New York: Plenum.

Nicol, T., Helmy, I. D. and Abou-Zikry, A. (1952). A histological explanation for the beneficial action of endocrine therapy in carcinoma of the prostate. *Br. med. J.* **40**, 166–72.

Nicol, T., Quantock, D. C. and Vernon-Roberts, B. (1966). Stimulation of phagocytosis in relation to the mechanism of action of adjuvants. *Nature, Lond.* **209**, 1142–3.

Nicol, T., Quantock, D. C. and Vernon-Roberts, B. (1967). The effects of steroid hormones on local and general reticuloendothelial activity: relation of steroid structure to function. In *The Reticuloendothelial System and Atherosclerosis* (eds. N. R. DiLuzio and R. Paoletti), pp. 221–42. New York: Plenum.

Nicol, T. and Vernon-Roberts, B. (1965). The influence of the estrus cycle, pregnancy and ovariectomy on RES activity. *J. Reticuloendothel. Soc.* **2**, 15–29.

Nicol, T., Vernon-Roberts, B. and Quantock, D. C. (1965*a*). Protective effect of oestrogens against the toxic decomposition products of tribromoethanol. *Nature, Lond.* **208**, 1098–9.

Nicol, T., Vernon-Roberts, B. and Quantock, D. C. (1965*b*). The influence of various hormones on the reticulo-endothelial system: endocrine control of body defence. *J. Endocr.* **33**, 365–83.

Nicol, T., Vernon-Roberts, B. and Quantock, D. C. (1966*a*). The effects of oestrogen:androgen interaction on the reticulo-endothelial system and reproductive tract. *J. Endocr.* **34**, 163–78.

Nicol, T., Vernon-Roberts, B. and Quantock, D. C. (1966*b*). The effects of various anti-oestrogenic compounds on the reticulo-endothelial system and reproductive tract in the ovariectomized mouse. *J. Endocr.* **34**, 377–86.

Nicol, T., Vernon-Roberts, B. and Quantock, D. C. (1966*c*). Oestrogenic and anti-oestrogenic effects of oestriol, 16-epi-oestriol, 2-methoxyoestrone and 2-hydroxy-oestradiol-17β on the reticulo-endothelial system and reproductive tract. *J. Endocr.* **35**, 119–20.

Nicol, T., Vernon-Roberts, B. and Quantock, D. C. (1967). The effect of testosterone and progesterone on the response of the reticulo-endothelial system and reproductive tract to oestrogen in the male mouse. *J. Endocr.* **37**, 17–21.

Nicol, T. and Zikry, A. A. (1952). Effect of oestradiol benzoate and orchidectomy on the toxicity of trypan blue. *Nature, Lond.* **170**, 239–40.

Normann, S. J. and Benditt, E. P. (1965). Function of the reticuloendothelial system. II. Participation of a serum factor in carbon clearance. *J. exp. Med.* **122**, 709–19.

North, R. J. (1966*a*). The localization by electron microscopy of nucleoside phosphatase activity in guinea-pig phagocytic cells. *J. Ultrastruct. Res.* **16**, 83–95.

North, R. J. (1966*b*). The localization by electron microscopy of acid phosphatase activity in guinea-pig macrophages. *J. Ultrastruct. Res.* **16**, 96–108.

North, R. J. (1969*a*). Cellular kinetics associated with the development of acquired cellular resistance. *J. exp. Med.* **130**, 299–314.

North, R. J. (1969*b*). The mitotic potential of fixed phagocytes in the liver as revealed during the development of cellular immunity. *J. exp. Med.* **130**, 315–26.

Norton, W. L. and Ziff, M. (1966). Electron microscopic observations on the rheumatoid synovial membrane. *Arthritis Rheum.* **9**, 589–610.

Nossal, G. J. V., Abbot, A. and Mitchell, J. (1968). Antigens in immunity. XIV. Electron microscopic radioautographic studies of antigen capture in the lymph node medulla. *J. exp. Med.* **127**, 263–76.

Nossal, G. J. V., Abbot, A., Mitchell, J. and Lummus, Z. (1968). Antigens in immunity. XV. Ultrastructural features of antigen capture in primary and secondary lymphoid follicles. *J. exp. Med.* **127**, 277–90.

Nossal, G. J. V., Ada, G. L. and Austin, C. M. (1964*a*). Antigens in immunity. IV. Cellular localization of [125]I- and ^{131}I-labelled flagella in lymph nodes. *Aust. J. exp. Biol. med. Sci.* **42**, 311–30.

Nossal, G. J. V., Ada, G. L. and Austin, C. M. (1964*b*). Antigens in immunity. V. The ability of cells in lymphoid follicles to recognize foreignness. *Aust. J. exp. Biol. med. Sci.* **42**, 331–46.

Nossal, G. J. V., Ada, G. L. and Austin, C. M. (1965). Antigens in immunity. IX. antigen content of single antibody-forming cells. *J. exp. Med.* **121**, 945–54.

J. V., Ada, G. L. and Pye, J. (1964). Antigens in immunity. III. Distribunated antigens following injection into rats via the hind footpads. *Aust. d. Sci.* **42**, 295–310.

Nossal, G. J. V., Austin, C. M., Pye, J. and Mitchell, J. (1966). Antigens in immunity. XII. Antigen trapping in the spleen. *Int. Archs Allergy appl. Immun.* **29**, 368–83.

Nossal, G. J. V., Cunningham, A., Mitchell, G. F. and Miller, J. F. A. P. (1968). Cell to cell interaction in the immune response. III. Chromosomal marker analysis of single antibody-forming cells in reconstituted, irradiated, or thymectomized mice. *J. exp. Med.* **128**, 839–50.

Novikoff, A. B. (1961). Lysosomes and related particles. In *The Cell*, Vol. II (eds. J. Brachet and A. E. Mirsky), pp. 423–88. New York & London: Academic Press.

Old, L. J., Boyse, E. A., Bennett, B. and Lilly, F. (1963). Peritoneal cells as an immune population in transplantation studies. In *Cell-bound Antibodies* (eds. B. Amos and H. Koprowski), pp. 89–98. Philadelphia: Wistar Institute.

Old, L. J., Boyse, E. A., Clarke, D. A. and Carswell, E. A. (1962). Antigenic properties of chemically induced tumours. *Ann. N.Y. Acad. Sci.* **101**, 80–106.

Old, L. J., Clarke, D. A., Benacerraf, B. and Goldsmith, M. (1960). The reticuloendothelial system and the neoplastic process. *Ann. N.Y. Acad. Sci.* **88**, 264–80.

Oren, R., Farnham, A. E., Saito, K., Milofsky, E. and Karnovsky, M. L. (1963). Metabolic patterns in three types of phagocytosing cells. *J. biophys. biochem. Cytol.* **17**, 487–501.

Osmond, D. G. and Everett, N. B. (1964). Radioautographic studies on the bone marrow lymphocytes *in vivo* in diffusion chamber cultures. *Blood*, **23**, 1–17.

Page, A. R. and Good, R. A. (1958). A clinical and experimental study of the function of neutrophils in the inflammatory response. *Am. J. Path.* **34**, 645–70.

Papanicolaou, G. N. (1953). Observations on the origin and specific function of the histiocytes in the female genital tract. *Fertil. Steril.* **4**, 472–78.

Parker, H. G. and Finney, C. R. (1960). Latent period in the induction of reticuloendothelial blockade. *Am. J. Physiol.* **198**, 916–20.

Parrott, D. M. V. (1967). The response of draining lymph nodes to immunological stimulation in intact and thymectomized animals. *J. clin. Path.* **20**, 456–65.

Patek, P. R. and Bernick, S. (1960). Time sequence studies of the reticuloendothelial cell response to foreign particles. *Anat. Rec.* **138**, 27–37.

Patek, P. R., Bernick, S. and de Mignard, V. A. (1967). Arteriopathy induced by reticuloendothelial blockade. In *The Reticuloendothelial System and Atherosclerosis* (eds. N. R. DiLuzio and R. Paoletti), pp. 413–25. New York: Plenum.

Patek, P. R., Bernick, S. and Frankel, H. H. (1961). Arterial lesions in rats by reticuloendothelial blocking agents. *Archs Path.* **72**, 70–8.

Patterson, R. and Suszko, I. M. (1966). Passive immune elimination and *in vitro* phagocytosis of antigen–antibody complexes in relation to specific origin of antibody. *J. Immun.* **97**, 138–49.

Pavillard, E. R. J. and Rowley, D. (1962). A comparison of the phagocytic ability of guinea-pig alveolar and mouse peritoneal macrophages. *Aust. J. exp. Biol. med. Sci.* **40**, 207–14.

Perkins, E. H. and Makinodan, T. (1965). The suppressive role of mouse peritoneal phagocytes in agglutinin responses. *J. Immun.* **94**, 765–77.

Perkins, E. H., Nettesheim, P. and Morita, T. (1966). Radioresistance of the engulfing and degradative capacities of peritoneal phagocytes to kiloroentgen X-ray doses. *J. Reticuloendothel. Soc.* **3**, 71–82.

Perkins, E. H., Nettesheim, P., Morita, T. and Walburg, H. E. (1967). The engulfing potential of peritoneal phagocytes of conventional and germfree mice. In *The Reticuloendothelial System and Atherosclerosis* (eds. N. R. DiLuzio and R. Paoletti), pp. 175–87. New York: Plenum.

Pernis, B., Bairati, A. and Milanesi, S. (1966). Cellular and humoral reactions to Freund's adjuvant in guinea-pigs. *Pathologia Microbiol.* **29**, 837–53.

Pethica, B. A. (1961). The physical chemistry of cell adhesion. *Expl Cell Res.* Suppl. **8**, 123–40.

Phillips, M. E. and Thorbecke, G. J. (1966). Studies on the serum proteins of chimeras. I. Identification and study of the site of origin of donor type serum proteins in adult rat-into-mouse chimeras. *Int. Archs. Allergy appl. Immun.* **29**, 553–67.

Pinkett, M. O., Cowdrey, C. R. and Nowell, P. C. (1966). Mixed haemopoietic and pulmonary origin of 'alveolar macrophages' as demonstrated by chromosome markers. *Am. J. Path.* **48**, 859–67.

Plager, J. E. and Samuels, L. T. (1954). The conversion of progesterone to 17-hydroxy-11-desoxycorticosterone by fractionated beef adrenal homogenates. *J. biol. Chem.* **211**, 21–9.

Poole, J. C. F. (1966). Phagocytosis of platelets by monocytes in organizing arterial thrombi. *Q. Jl exp. Physiol.* **51**, 54–9.

Poole, J. C. F. and Florey, H. W. (1958). Changes in the endothelium of the aorta and the behaviour of macrophages in experimental atheroma of rabbits. *J. Path. Bact.* **75**, 245–51.

Porter, K. A. (1967). The immune response to tissue transplants. In *Modern Trends in Pathology*, Vol. II (ed. T. Crawford), pp. 102–39. London: Butterworths.

Porter, K. A. and Calne, R. Y. (1960). The origin of the infiltrating cells in skin and kidney homografts. *Transplantn Bull.* **26**, 458–64.

Prehn, R. J. and Main, J. M. (1957). Immunity to methylcholanthrene-induced sarcomas. *J. natn. Cancer Inst.* **18**, 769–78.

Quantock, D. C. (1969). The effects of oestrogens, cytotoxic compounds and irradiation on human phagocytic activity. *M.D. Thesis, University of London.*

Rabinovitch, M. (1967a). Attachment of modified erythrocytes to phagocytic cells in absence of serum. *Proc. Soc. exp. Biol. Med.* **124**, 396–9.

Rabinovitch, M. (1967b). The dissociation of the attachment and ingestion phases of phagocytosis by macrophages. *Expl Cell Res.* **46**, 19–28.

Reade, P. C. and Jenkin, C. R. (1965). The functional development of the reticulo-endothelial system. I. The uptake of intravenously injected particles by foetal rats. *Immunology*, **9**, 53–60.

Rebuck, J. W., Boyd, C. B. and Riddle, J. M. (1960). Skin windows and the action of the reticuloendothelial system in man. *Ann. N.Y. Acad. Sci.* **88**, 30–42.

Rebuck, J. W., Coffman, H. I., Bluhm, G. B. and Barth, C. L. (1964). A structural study of reticulum cell and monocyte production with quantitation of lymphocyte modulation of nonmultiplicative types to histiocytes. *Ann. N.Y. Acad. Sci.* **113**, 595–611.

Rebuck, J. W. and Crowley, J. H. (1955). A method for studying leucocytic function in vivo. *Ann. N.Y. Acad. Sci.* **59**, 757–805.

Rebuck, J. W., Monto, R. W., Monaghan, E. A. and Riddle, J. M. (1958). Potentialities of the lymphocyte with an additional reference to its dysfunction in Hodgkin's disease. *Ann. N.Y. Acad. Sci.* **73**, 8–39.

Rebuck, J. W., Whitehouse, F. W. and Noonan, S. M. (1967). A major fault in diabetic inflammation: failure of leucocytic glycogen transfer to histiocytes. In *The Reticuloendothelial System and Atherosclerosis* (eds. N. R. DiLuzio and R. Paoletti), pp. 369–81. New York: Plenum.

Rhodes, J. M. (1964). *In vitro* studies on the fate of antigen. IV. The fate of soluble and precipitated antigen–antibody complexes in the presence of guinea pig exudate cells, correlated with their skin reactivity. *Int. Archs Allergy appl. Immun.* **25**, 225–41.

Robineaux, R. and Pinet, J. (1960). An *in vitro* study of some mechanisms of antigen uptake by cells. In *Cellular Aspects of Immunity* (Ciba Foundation Symposium) (eds. G. E. W. Wolstenholme and M. O'Connor), pp. 5–40. London: Churchill.

Robinson, H. J. (1953). The role of the adrenal glands in infection and intoxication. In *The Suprarenal Cortex* (ed. J. M. Yoffey), pp. 105–24. London: Butterworths.

Roelants, G. E. and Goodman, J. W. (1969). The chemical nature of macrophage RNA–antigen complexes and their relevance to immune induction. *J. exp. Med.* **130**, 557–74.

Roitt, I. M. and Doniach, D. (1967). Delayed hypersensitivity in auto-immune disease. *Br. med. Bull.* **23**, 66–71.

Roitt, I. M., Jones, H. E. H. and Doniach, D. (1961). Mechanisms of tissue damage in human and experimental auto-immune thyroiditis. In *Mechanism of Cell and Tissue Damage produced by Immune Reactions* (2nd International Symposium on Immunopathology) (eds. P. Grabar and P. A. Miescher), pp. 174–83. Basel: Schwabe.

Rosenau, W. and Morton, D. L. (1966). Tumour-specific inhibition of growth of methylcholanthrene-induced sarcomas *in vivo* and *in vitro* by sensitized isologous lymphoid cells. *J. natn. Cancer Inst.* **36**, 825–36.

Rosenberg, A. and Chargaff, E. (1958). Reinvestigation of cerebroside deposited in Gaucher's diseases. *J. biol. Chem.* **233**, 1323–6.

Rosenman, R. H., Breall, W. and Friedman, M. (1960). A study of hepatic reticuloendothelial cell function in experimental nephrotic hyperlipemia and hypercholesterolemia. In *Reticulo-endothelial Structure and Function* (ed. J. H. Heller), pp. 417–29. New York: Ronald.

Roser, B. (1965). The distribution of intravenously injected peritoneal macrophages in the mouse. *Aust. J. exp. Biol. med. Sci.* **43**, 553–62.

Roser, B. (1968). The distribution of intravenously injected Kupffer cells in the mouse. *J. Reticuloendothel. Soc.* **5**, 455–71.

Rowley, D. (1958). Bactericidal activity of macrophages *in vitro* against *Escherichia coli*. *Nature, Lond.* **181**, 1738–9.

Rowley, D. and Leuchtenberger, C. (1964). Antigen-stimulated desoxyribonucleic acid synthesis *in vitro* by sensitized mouse macrophages. *Lancet*, **2**, 734–5.

Rowley, D., Turner, K. J. and Jenkin, C. R. (1964). The basis for immunity to mouse typhoid. 3. Cell-bound antibody. *Aust. J. exp. Biol. med. Sci.* **42**, 237–48.

Russell, P. and Roser, B. (1966). The distribution and behaviour of intravenously injected alveolar macrophages in the mouse. *Aust. J. exp. Biol. med. Sci.* **44**, 629–38.

Ryan, G. B. and Spector, W. G. (1970). Macrophage turnover in inflamed connective tissues. *Proc. R. Soc. B.* **175**, 269–92.

Salky, N. K., DiLuzio, N. R., Levin, N. R. and Goldsmith, H. S. (1967). Phagocytic activity of the reticulo-endothelial system in neoplastic disease. *J. Lab. clin. Med.* **70**, 393–403.

Sbarra, A. J. and Karnovsky, M. L. (1959). The biochemical basis of phagocytosis. I. Metabolic changes during the ingestion of particles by polymorphonuclear leukocytes. *J. biol. Chem.* **234**, 1355–62.

Sbarra, A. J., Shirley, W. and Bardawil, W. A. (1962). 'Piggy-back' phagocytosis. *Nature, Lond.* **194**, 255–6.

Schaffner, F. and Popper, H. (1962). A phagocytic and protein-forming mesenchymal cell in human cirrhosis. *Nature, Lond.* **196**, 684–5.

Schapiro, R. L., MacIntyre, W. J. and Schapiro, D. I. (1966). The effect of homologous and heterologous carrier on the clearance of colloidal material by the reticulo-endothelial system. *J. Lab. clin. Med.* **68**, 286–99.

Schoenberg, M. D., Mumaw, V. R., Moore, R. D. and Weisberger, A. S. (1964). Cytoplasmic interaction between macrophages and lymphocytic cells in antibody synthesis. *Science, N.Y.* **143**, 964–5.

Schooley, J. C., Kelly, L. S., Dobson, E. L., Finney, C. R., Havens, V. W. and Cantor, L. N. (1965). Reticuloendothelial activity in neonatally thymectomized mice and irradiated mice thymectomized in adult life. *J. Reticuloendothel. Soc.* **2**, 396–405.

Schwartz, R. S. and Beldotti, L. (1965). Malignant lymphomas following allogenic disease: transition from an immunological to a neoplastic disorder. *Science, N.Y.* **149**, 1511–14.

Scothorne, R. J. and McGregor, I. A. (1955). Cellular changes in lymph nodes and spleen following skin homografting in the rabbit. *J. Anat.* **89**, 283–92.

Sharp, J. A. (1968). The role of macrophages in the induction of the immune response. *Int. Rev. gen. exp. Zool.* **3**, 117–70.

Sharp, J. A. and Burwell, R. G. (1960). Interaction ('Peripolesis') of macrophages and lymphocytes after skin homografting or challenge with soluble antigens. *Nature, Lond.* **188**, 474–5.

Sheagren, J. N., Barth, R. F., Edelin, J. B. and Malmgren, R. A. (1969). Reticuloendothelial blockade produced by antilymphocyte serum. *Lancet.* **2**, 297–8.

Sheagren, J. N., Block, J. B. and Wolff, S. M. (1967). Reticulo-endothelial system phagocytic function in patients with Hodgkin's disease. *J. clin. Invest.* **16**, 855–62.

Shearing, S. P., Comerford, F. R., and Cohen, A. S. (1965). Effect of an amyloid inducing regimen on phagocytosis of carbon particles. *Proc. Soc. exp. Biol. Med.* **119**, 673–6.

Shelton, E. and Rice, M. E. (1959). Growth of normal peritoneal cells in diffusion chambers: a study in cell modulation. *Am. J. Anat.* **105**, 281–342.

Shortman, K., Diener, E., Russell, P. and Armstrong, W. D. (1970). The role of nonlymphoid accessory cells in the immune response to different antigens. *J. exp. Med.* **131**, 461–82.

Silverstein, S. C. and Dales, S. (1968). The penetration of reovirus RNA and initiation of its genetic function in L-strain fibroblasts. *J. biophys. biochem. Cytol.* **36**, 197–230.

Simon, R. C., Still, W. J. S. and O'Neal, R. M. (1961). The circulating lipophages and experimental atherosclerosis. *J. Atheroscler. Res.* **1**, 395–400.

Šljivić, V. S. (1970*a*). Radiation and the phagocytic function of the reticuloendothelial system. I. Enhancement of RES function in X-irradiated mice. *Br. J. exp. Path.* **51**, 130–9.

Šljivić, V. S. (1970*b*). Radiation and the phagocytic function of the reticuloendothelial system. II. Mechanism of RES stimulation after irradiation. *Br. J. exp. Path.* **51**, 140–8.

Smetana, H. (1926). The relation of the reticulo-endothelial system to the formation of amyloid. *J. exp. Med.* **45**, 619–29.

Snell, J. F. (1960*a*). The reticuloendothelial system: I. Chemical methods of stimulation of the reticuloendothelial system. *Ann. N.Y. Acad. Sci.* **88**, 56–77.

Snell, J. F. (1960*b*). Relationship of chromium phosphate clearance rates to resistance: I. The effects of some corticosteroids on blood clearance rates in mice. In *Reticulo-endothelial Structure and Function* (ed. J. H. Heller), pp. 321–32. New York: Ronald.

Sorensen, G. D. (1960). An electron microscopic study of popliteal lymph nodes from rabbits. *Am. J. Anat.* **107**, 73–96.

Sorensen, G. D., Heefner, W. A. and Kirkpatrick, J. B. (1964). Experimental amyloidosis. II. Light and electron microscopic observations of liver. *Am. J. Path.* **44**, 629–44.

Sorkin, E., Borel, J. F. and Stecher, V. J. (1970). Chemotaxis of mononuclear and polymorphonuclear phagocytes. In *Mononuclear Phagocytes* (ed. R. Van Furth), pp. 397–418. Oxford: Blackwell.

Sorkin, E. and Boyden, S. V. (1959). Studies on the fate of antigens *in vitro*. *J. Immun.* **82**, 332–9.

Soulsby, E. J. L. (1962). Antigen–antibody reactions in helminth infections. *Adv. Immunol.* **2**, 265–308.

Southam, C. M., Moore, A. E. and Rhoads, C. P. (1957). Homotransplantation of human cell lines. *Science, N.Y.* **125**, 158–60.

Spector, W. G. (1967). Histology of allergic inflammation. *Br. med. Bull.* **23**, 35–8.

Spector, W. G. (1969). The granulomatous inflammatory exudate. *Int. Rev. exp. Pathol.* **8**, 1–55.

Spector, W. G. and Coote, E. (1965). Differentially labelled blood cells in the reaction to paraffin oil. *J. Path. Bact.* **90**, 589–98.

Spector, W. G. and Lykke, A. W. J. (1966). The cellular evolution of inflammatory granulomata. *J. Path. Bact.* **92**, 163–7.

Spector, W. G., Walters, M. N-I. and Willoughby, D. A. (1967). A quantitative study of leucocytic emigration in chronic inflammatory granulomata. *J. Path. Bact.* **93**, 101–8.

Spector, W. G. and Willoughby, D. A. (1968). The origin of mononuclear cells in chronic inflammation and tuberculin reactions in the rat. *J. Path. Bact.* **96**, 389–99.

Spiegelberg, H. L., Miescher, P. A. and Benacerraf, B. (1963). Studies on the role of complement in the immune clearance of *Escherichia coli* and rat erythrocytes by the reticuloendothelial system in mice. *J. Immun.* **90**, 751–9.

Spraragen, S. C., Bond, V. P. and Dahl, L. K. (1962). Role of hyperplasia in vascular lesions of cholesterol-fed rabbits studied with thymidine-H^3 autoradiography. *Circulation Res.* **11**, 329–36.

Stähelin, H., Suter, E. and Karnovsky, M. L. (1956). Studies on the interaction between phagocytes and tubercle bacilli. I. Observations on the metabolism of guinea pig leucocytes and the influence of phagocytosis. *J. exp. Med.* **104**, 121–36.

Stecher, V. J. and Thorbecke, G. J. (1967). Sites of synthesis of serum proteins. I. Serum proteins produced by macrophages *in vitro*. *J. Immun.* **99**, 643–52.

Stern, K. (1960). The reticulo-endothelial system and neoplasia. In *Reticulo-endothelial Structure and Function* (ed. J. H. Heller), pp. 233–58. New York: Ronald.

Stiffel, C., Biozzi, G., Mouton, D., Bouthillier, Y. and Decreusefond, C. (1964). Studies on phagocytosis of bacteria by the reticulo-endothelial system in a strain of mice lacking hemolytic complement. *J. Immun.* **93**, 246–9.

Strober, S. and Gowans, J. L. (1965). The role of lymphocytes in the sensitization of rats to renal homografts. *J. exp. Med.* **122**, 347–60.

Stuart, A. E. (1963). Structural and functional effects of lipids on the reticuloendothelial system. In *Rôle du Système Réticulo-endothélial dans l'immunité antibactérienne et antitumorale* (ed. B. N. Halpern), pp. 129–42. Paris: CRNS.

Stuart, A. E. (1970). *The Reticulo-endothelial System.* Edinburgh & London: Livingstone.

Stuart, A. E. and Cooper, G. N. (1962). Susceptibility of mice to bacterial endotoxin after modification of reticulo-endothelial function by simple lipids. *J. Path. Bact.* **83**, 245–54.

Stuart, A. E. and Cumming, R. A. (1967). A biological test for injury to the human red cell. *Vox. Sang.* **13**, 270–80.

Sutton, J. S. and Weiss, L. (1966). Transformation of monocytes in tissue culture into macrophages, epithelioid cells, and multinucleate giant cells. An electron microscope study. *J. Biophys. biochem. Cytol.* **28**, 303–32.

Szakel, A. K. and Hanna, M. G. (1968). The ultrastructure of antigen localization and virus-like particles in mouse spleen germinal centers. *J. exp. molec. Path.* **8**, 75–89.

Szenberg, A. and Warner, N. L. (1967). The role of antibody in delayed hypersensitivity. *Br. med. Bull.* **23**, 30–4.

Tanaka, M. (1961). Electron microscopic studies on the mechanism of vital stain, as compared with that of phagocytosis. *Proc. Jap. Soc. RES.* **1**, 44–60.

Teilum, G. (1952). Cortisone–ascorbic acid interaction and pathogenesis of amyloidosis; mechanism of action of cortisone on mesenchymal tissues. *Ann. rheum. Dis.* **11**, 119–35.

Teilum, G. (1954). Studies on pathogenesis of amyloidosis; effect of nitrogen mustard in inducing amyloidosis. *J. Lab. clin. Med.* **43**, 367–74.

Thomas, L. (1957). The role of the reticulo-endothelial system in the reaction to endotoxins. In *Physiopathology of the Reticulo-endothelial System* (eds. B. N. Halpern, B. Benacerraf and J. F. Delafresnaye), pp. 226–43. Oxford: Blackwell.

Trowell, O. A. (1958). The lymphocyte. *Int. Rev. Cytol.* **7**, 235–94.

Turk, J. L. (1967). Cytology of the induction of hypersensitivity. *Br. med. Bull.* **23**, 3–8.

Turk, J. L., Heather, C. J. and Diengdoh, J. V. (1966). A histochemical analysis of mononuclear cell infiltrates of the skin with particular reference to delayed hypersensitivity in the guinea pig. *Int. Archs Allergy appl. Immun.* **29**, 278–89.

Turk, J. L. and Polak, L. (1967). Studies on the origin and reactive ability *in vivo* of peritoneal exudate cells in delayed hypersensitivity. *Int. Archs Allergy appl. Immun.* **31**, 403–16.

Turk, J. L., Rudner, E. J. and Heather, C. J. (1966). A histochemical analysis of mononuclear cell infiltration of the skin. II. Delayed hypersensitivity in the human. *Int. Archs Allergy appl. Immun.* **30**, 248–56.

Uhr, J. W. (1965). Passive sensitization of lymphocytes and macrophages by antigen–antibody complexes. *Proc. natn. Acad. Sci. U.S.A.* **54**, 1599–1606.

Uhr, J. W. and Weissmann, G. (1965). Intracellular distribution and degradation of bacteriophage in mammalian tissue. *J. Immun.* **94**, 544–50.

Unanue, E. R. and Askonas, B. A. (1968). Persistence of immunogenicity of antigen after uptake by macrophages. *J. exp. Med.* **127**, 915–26.

Unanue, E. R. and Cerottini, J-R. (1970). The immunogenicity of antigen bound to the plasma membrane of macrophages. *J. exp. Med.* **131**, 711–25.

Van Furth, R. (1970). Origin and kinetics of monocytes and macrophages. *Seminars Hematol.* **7**, 125–41.

Van Furth, R. and Cohn, Z. A. (1968). The origin and kinetics of mononuclear phagocytes. *J. exp. Med.* **128**, 415–35.

Van Furth, R. and Diesselhoff-Den Dulk, M. M. C. (1970). The kinetics of promonocytes and monocytes in the bone marrow. *J. exp. Med.* **132**, 813–28.

Van Furth, R., Hirsch, J. G. and Fedorko, M. E. (1970). Morphology and peroxidase cytochemistry of mouse promonocytes, monocytes, and macrophages. *J. exp. Med.* **132**, 794–812.

Vaughan, R. B. (1965a). The discriminative behaviour of rabbit phagocytes. *Br. J. exp. Path.* **46**, 71–81.

Vaughan, R. B. (1965*b*). Interactions of macrophages and erythrocytes: some further experiments. *Immunology*, **8**, 245–50.

Vaughan, R. and Boyden, S. V. (1964). Interactions of macrophages and erythrocytes. *Immunology*, **7**, 118–26.

Vernon-Roberts, B. (1969*a*). Lymphocyte to macrophage transformation in the peritoneal cavity preceding the mobilization of peritoneal macrophages to inflamed areas. *Nature, Lond.* **222**, 1286–8.

Vernon-Roberts, B. (1969*b*). The effects of steroid hormones on macrophage activity. *Int. Rev. Cytol.* **25**, 131–59.

Vigliani, E. C. and Pernis, B. (1958). Immunological factors in the pathogenesis of the hyaline tissue of silicosis. *Br. J. Ind. Med.* **15**, 8–14.

Virolainen, M. (1968). Hematopoietic origin of macrophages as studied by chromosome markers in mice. *J. exp. Med.* **127**, 943–53.

Volkman, A. (1966). The origin and turnover of mononuclear cells in peritoneal exudates in rats. *J. exp. Med.* **124**, 241–53.

Volkman, A. and Collins, F. M. (1968). Recovery of delayed-type hypersensitivity in mice following suppressive doses of *X*-radiation. *J. Immun.* **101**, 846–59.

Volkman, A. and Gowans, J. L. (1965*a*). The production of macrophages in the rat. *Br. J. exp. Path.* **46**, 50–61.

Volkman, A. and Gowans, J. L. (1965*b*). The origin of macrophages from bone marrow in the rat. *Br. J. exp. Path.* **46**, 62–70.

von Recklinghausen, F. (1863). Ueber Eiter- und Bindegewebsköperchen. *Virchows Arch. path. Anat. Physiol.* **28**, 157.

Waddell, W. R., Geyer, R. P., Clarke, E. and Stare, F. J. (1954). Function of the reticuloendothelial system in removal of emulsified fat from blood. *Am. J. Physiol.* **177**, 90–4.

Wagner, H. N. and Ilio, M. (1964). Studies of the reticuloendothelial system (RES). III. Blockade of the RES in man. *J. clin. Invest.* **43**, 1525–32.

Wagner, H. N., Ilio, M. and Hornick, R. B. (1963). Studies of the reticuloendothelial system (RES). II. Changes in the phagocytic activity of the RES in patients with certain infections. *J. clin. Invest.* **42**, 427–34.

Wagner, H. N., Migita, T. and Solomon, N. (1966). Effect of age on reticuloendothelial function in man. *J. Geront.* **21**, 57–62.

Wagner, R. R. and Smith, T. J. (1968). On the apparent heterogeneity of rabbit interferon. In *Interferon* (Ciba Foundation Symposium) (eds. G. E. W. Wolstenholme and M. O'Connor), pp. 95–109. London: Churchill.

Walford, R. L. and Hildemann, W. H. (1965). Life span and lymphoma-incidence of mice injected at birth with spleen cells across a weak histoincompatibility locus. *Am. J. Pathol.* **47**, 713–19.

Ward, P. A., Remold, H. G. and David, J. R. (1969). Leukotactic factor produced by sensitized lymphocytes. *Science, N.Y.* **163**, 1079–81.

Wardlaw, A. C. and Howard, J. G. (1959). A comparative survey of the phagocytosis of different species of bacteria by Kupffer cells. *Br. J. exp. Path.* **40**, 113–17.

Wartman, W. B. (1959). Sinus cell hyperplasia of lymph nodes regional to adenocarcinoma of the breast and colon. *Br. J. Cancer*, **13**, 389–97.

Watts, J. H. and Harris, H. (1959). Turnover of nucleic acids in a non-multiplying animal cell. *Biochem. J.* **72**, 147–53.

Weiser, R. S., Granger, G. A., Brown, W., Baker, P., Jutila, J. and Holmes, B. (1965). Production of acute allogeneic disease in mice. *Transplantation*, **3**, 10–21.

Weiss, L. and Fawcett, D. W. (1953). Cytochemical observations on chicken monocytes, macrophages and giant cells in tissue culture. *J. Histochem. Cytochem.* **1**, 47–65.

Weissman, G. and Thomas, L. (1963). Studies on lysosomes. II. The effect of cortisone on the release of acid hydrolases from a large granule fraction of rabbit liver induced by an excess of Vitamin A. *J. clin. Invest.* **42**, 661–9.

White, R. G. (1963). Functional recognition of immunologically competent cells by means of the fluorescent antibody technique. In *The Immunologically Competent Cell* (Ciba Foundation: Study Group No. 16) (eds. G. E. W. Wolstenholme and J. Knight), pp. 6–16. London: Churchill.

White, R. G., French, V. I. and Stark, J. M. (1970). A study of the localization of a protein antigen in the chicken spleen and its relation to the formation of germinal centres. *J. med. Microbiol.* **3**, 65–81.

Whitelaw, D. M. (1966). The intravascular lifespan of monocytes. *Blood*, **28**, 455–64.

Wiener, J. (1967). Fine structural aspects of reticuloendothelial blockade. In *The Reticuloendothelial System and Atherosclerosis* (eds. N. R. DiLuzio and R. Paoletti), pp. 85–97. New York: Plenum.

Wiener, J. and Spiro, D. (1962). Electron microscope studies in experimental thrombosis. *Exp. Mol. Pathol.* **1**, 554–72.

Wiener, J., Spiro, D. and Russell, P. (1964). An electron microscopic study of the homograft reaction. *Am. J. Path.* **44**, 319–47.

Wilkins, D. J. and Myers, P. A. (1966). Studies on the relationship between electrophoretic properties of colloids and their blood clearance and organ distribution in the rat. *Br. J. exp. Path.* **47**, 568–76.

Wisse, E. (1970). An electron microscopic study of the fenestrated endothelial lining of rat liver sinusoids. *J. Ultrastruct. Res.* **31**, 125–50.

Wood, W. B., Smith, M. R. and Watson, B. (1958). Studies on the mechanism of recovery in pneumococcal pneumonia. IV. The mechanism of phagocytosis in the absence of antibody. *J. exp. Med.* **84**, 387–402.

Yamakawa, T. (1967). Biochemistry of lipoidoses. *Proc. Jap. Soc. RES.* **7**, 1–9.

Zilversmit, D. B., McCandless, E. L., Jordan, P. H., Henley, W. S. and Ackerman, R. F. (1961). The synthesis of phospholipids in human atheromatous lesions. *Circulation*, **23**, 370–5.

INDEX